Advance Praise for

Christian Theology in a Pluralistic Context

"Christian theology claims that revelation is wide-reaching; yet our constructive theology too often remains parochial. Steven L. Wiebe tellingly explores this inconsistency with regard to a doctrine of creation, and develops out of the biblical wisdom tradition a helpful methodological corrective. This book is an essential read for those wanting to engage our pluralistic age."

Robert K. Johnston, Professor of Theology and Culture, Fuller Theological Seminary

"Christian theology in the twenty-first century will be done in conversation with other religions' theologies. Steven L. Wiebe's work is an example of how that might be done. This work pushes us in the right direction. I highly recommend it."

Terry C. Muck, Professor of Missions and World Religions, Asbury Theological Seminary

Christian Theology in a Pluralistic Context

american
university
studies

Series VII
Theology and Religion

Vol. 255

PETER LANG
New York • Washington, D.C./Baltimore • Bern
Frankfurt am Main • Berlin • Brussels • Vienna • Oxford

Steven L. Wiebe

Christian Theology in a Pluralistic Context

A Methodological and Constructive Inquiry in the Doctrine of Creation

PETER LANG
New York • Washington, D.C./Baltimore • Bern
Frankfurt am Main • Berlin • Brussels • Vienna • Oxford

Library of Congress Cataloging-in-Publication Data

Wiebe, Steven L.
Christian theology in a pluralistic context: a methodological
and constructive inquiry in the doctrine of creation / Steven L. Wiebe.
p. cm. — (American university studies. VII, Theology and religion; v. 255)
Includes bibliographical references and index.
1. Creationism. 2. Cosmogony. 3. Postmodernism—Religious
aspects—Christianity. 4. Religious pluralism. I. Title.
BS651.W4265 231.7'65—dc22 2006022710
ISBN-13: 978-0-8204-8827-1
ISBN-10: 0-8204-8827-5
ISSN 0740-0446

Bibliographic information published by **Die Deutsche Bibliothek**.
Die Deutsche Bibliothek lists this publication in the "Deutsche
Nationalbibliografie"; detailed bibliographic data is available
on the Internet at http://dnb.ddb.de/.

The paper in this book meets the guidelines for permanence and durability
of the Committee on Production Guidelines for Book Longevity
of the Council of Library Resources.

29 Broadway, 18th floor, New York, NY 10006
www.peterlang.com

Printed in Germany

To Professor Thomas E. Schmidt,
who taught me how to think.

Contents

Acknowledgments

Like many courses of study, this work blossomed out of the serendipity that is known quite well to writers and researchers. One question led to another, and to another, and lo and behold, a dissertation appeared. Mind you, this was not a short process, but one labored at over many years, as I was working full time during this period. My advisor at Fuller Seminary, Rob Johnston particularly helped shape and mold both me and this work, even during dark editorial days. Steven Bouma-Prediger, Terry Muck, and Veli-Matti Karkkainen all read the manuscript and offered timely comments. Turning this dissertation into book form happened under the helpful eye of Brittany Schwartz. My wife, Jennifer, and our three children, Alexis, Michaela, and Owen have afforded me much more time writing than is to be expected in a busy household, and for that I am so grateful. But in the end, the bulk of the work was accomplished at a computer in a study that is both part of a family home and the nerve center for a thriving urban outreach. It is my hope that this book in some way reflects the "real world" that it was composed in.

CHAPTER ONE

Introduction & Context

Introduction

Any theologian approaching the area of the traditional doctrine of creation is met at the outset by the sheer difficulty of speaking of anything "in the beginning". Dietrich Bonhoeffer has written elegantly of the near impossibility of understanding what happened "in the beginning": "for we cannot speak of the beginning; where the beginning begins our thinking stops, it comes to an end.... In thinking of the beginning thinking collapses." [1] The reason for the difficulty for Bonhoeffer is, of course, that we exist "in the middle," and that the beginning is beyond the reach of our reason.

Undaunted by this chasm, Christian theologians have proceeded to outline the main contours of a doctrine of "the beginning." From a fairly early stage in Christian theology, first with Irenaeus and then Origen in the 3rd century, theologians have considered a statement about the beginning crucial to any theological corpus. And, after a bit of a hiatus brought about by higher criticism at the beginning of the twentieth century, the doctrine of creation has seen a recent surge of interest in the wake of Karl Barth's intense focus on it in his *Dogmatics*.

But we are now living in a new epistemological world, that of postmodernity, whose tenets include a commitment to a "localized" narrative, and an aversion to anything that would seem to be a "meta–story".[2] We are also living in a time of great global diversity and interaction (a fact not unrelated to the rise of postmodernity), and more importantly, a time of being conscious of that diversity. This, as we shall see, imbues our theological task with different hues than previous generations'. The issue of plurality (with regard to faith) in such a diverse world has risen quickly and with full force. And since the issues in a doctrine of creation are so crucial to an understanding of our world, the people in it, and most decisively, God, the absolute centrality of re–addressing the doctrine in light of our current era, and particularly in light of the glare of pluralism, is readily apparent. Is it possible to craft a pluralistic Christian doctrine of creation? Is a Christian doctrine of creation closed to a dialogue with other religious sources?

Exploring these issues will require us to address the concern of many theologians and critics of Christianity alike, and that is a tendency to focus on doctrines of redemption to the exclusion of other matters. Ludwig Feuerbach put it succinctly, "Nature, the world, has no value, no interest, for Christians. The Christian thinks only of himself and the salvation of his soul." [3] This interest has been seen not only culturally and within the context of ministry among Christians, but also theologically, as theologians have tended to concentrate more on the topic of redemption, and not, for example, on a doctrine of creation. And while Feuerbach's criticism of Christianity may not be totally accurate, still one has to negotiate the influence of a theologian such as Karl Barth whose doctrine of creation is so focused on future redemption and covenant that it cannot stand alone as a dialogue partner with other religions. Can a Christian doctrine of creation enter a discussion and debate with other religions without always having to resort to a doctrine of redemption as the final *coup de grace*? [4] I believe it can, and this is precisely what we will be attempting to do in this book – interacting with other religions in the area of creation.

What we are exploring, then, is how we might develop a doctrine of creation that responds to the worldwide multi–religious context that we are a part of today. Theologically speaking, there are questions that arise with our approach. Why would one want a doctrine of creation that is informed by material from outside the Christian tradition or by the perspectives of other religious traditions? Is it not simply enough to state a constructive self–consciously Christian doctrine? The answer to the latter question is that of course it is sufficient to remain in one's own tradition in creating theology, and that is what most theologians have done in the history of the Christian church (though, as we will see, many theologians have made use of philosophical sources). However, we will be arguing that our cultural experience is one of pluralism, and as such, our theologies should reflect that context. This does not mean that we should become "pluralists," as we shall see, but rather crafters of pluralistic theology. Postmodernity helps us here as well, in both encouraging the more established endeavor of creating theology that reflects one's tradition, while at the same time suggesting that being open to new streams of information from outside of one's context is central to offering one's tradition to the diverse world. We will have more to say about postmodernity in our 2nd chapter on epistemological rationales.

However, another concern arises as we discuss a pluralistic doctrine of creation, and that is, how does one go about doing it? Do we simply line up the different religions and pick and choose from each one according to our proclivities? Do we take the Christian story as normative, give a cursory examination to the other religions, but ultimately ignore those other stories in an attempt to be true to our own narrative? Or do we attempt to augment the Christian story with insights from other religions? These types of questions force us to consider how pluralistic theology gets done, and so a major area in this book will be a development of a methodology for how to do pluralistic theology. But before we delve into that, we must consider our pluralistic context in more detail.

Pluralistic Context

As we mentioned earlier, our context is now global. Television and the internet have made us acutely aware of happenings all around the world, and it is only fitting that this global, diverse context would have implications for theology. Our theological task in this era must take into account our "neighbors". And in this way, our theological task is somewhat different than that of earlier generations of theologians. As we will see in chapter four, much of the first 1800 years of the Christian tradition was focused on clearly defining the Christian faith. The early church had to differentiate itself from competing religions to establish an identity rooted in the work of Christ; the Reformation was concerned with refining a Protestant viewpoint over against Catholicism; Neo–orthodoxy attempted to preserve a sense of transcendence in an age of higher criticism. But in our era, we must contemplate the role of the Christian faith in light of our knowledge of, proximity to, and competition with other world religions. This is simply our *sitz im leben* as theologians today. And so, in the last thirty years or so, we have seen the rise of "pluralism" as an area in Christian theology, led by theologians such as John Hick, Paul Knitter, and Raimundo Panikkar. And though we have also moved into an era of postmodernity, which may cast some serious questions on the entire pluralist construct, the reality of our pluralistic context endures, albeit in a different form. The question of comparing and evaluating traditions is a pointed one, even in a postmodern era, as is seen in books such as Alasdair MacIntyre's *Three Rival Versions of Moral Enquiry*. Nevertheless, we must be clear about what we mean when we say "pluralism."

In their seminal book on pluralism, *The Myth of Christian Uniqueness* (1987), Hick & Knitter argue that "a pluralistic model represents a new turn – what might be called a 'paradigm shift' – in the efforts of Christian theologians, both past and present, to understand the world of other religions and Christianity's place in that world." [5] Paul Knitter then outlines two broad models that have dominated Christian attitudes towards other religions:

> [1] the 'conservative' *exclusivist* approach, which finds salvation only in Christ and little, if any, value elsewhere …

> [2] the 'liberal' *inclusivist* attitude, which recognizes the salvific richness of other faiths but then views this richness as the result of Christ's redemptive work and as having to be fulfilled in Christ …

And then there is a third position, the adoption of which amounts to the crossing of a "theological Rubicon" as Knitter says, a shift from both the exclusivist and the inclusivist approach toward a

> [3] a *pluralist* position – a move away from insistence on the superiority or finality of Christ and Christianity toward a recognition of the independent validity of other ways. [6]

While noting that there are many other ways of describing the differing pluralistic positions, in the main, we will follow this threefold typology. [7]

There are three motivating factors in crossing this Rubicon – "bridges", according to Knitter. The first is *relativity*, and the "difficulty, if not impossibility, of judging the truth claims of another culture or religion on the basis of one's own." [8] Advocating one religion as superior to another is simply an untenable position, according to John Hick, if we examine the "facts":

> It seems impossible to make the global judgment that any one religious tradition has contributed more good or less evil, or a more favorable balance of good and evil,

> than the others ...We may well judge that in some respects, or in some periods or regions, the fruits of one tradition are better than, whereas in other respects or periods or regions inferior to, those of another. But as vast complex totalities, the world traditions seem to be more or less on a par with each other. None can be singled out as manifestly superior. [9]

The second bridge is *mystery*, by which is meant that the "object or content of authentic religious experience is infinite .... The infinity and ineffability of God–Mystery demands religious pluralism and forbids any one religion from having the 'only' or 'final' word." [10] The third bridge by which we cross over to pluralism is the recognition of the need for *justice* to confront the sufferings of humanity. If we are serious about ending injustice in the world, we must do so first by ending patterns of conflicting religions and come together in a pluralistic theology of religion. As Hans Küng puts it, "no world peace without peace among the religions, no peace among the religions without dialogue between the religions, and no dialogue between the religions without accurate knowledge of one another." [11]

Raimundo Panikkar brings the imagery of pluralism to life with an analogy, comparing the three theological positions of pluralism with rivers. [12] Exclusivism is represented by the *Jordan* river, with its roots in the Abrahamic covenant and Jesus' ministry. Only the Jordan can be the chosen river, in this perspective. The *Tiber* river denotes Christianity and Christendom, with its successful rise as a dominant world power in the last twenty centuries. The Tiber symbolizes the West and its ascendancy – sort of an "all roads lead to Rome" type of mentality – and would represent an inclusivist perspective. The *Ganges* is Panikkar's chosen symbol for pluralism, with its many sources, disappearing into innumerable tributaries. This Ganges stands "not just for Hinduism, Buddhism, Jainism, Sikhism, and primordial religions, but for all other traditions of Asia, Africa, and Oceania." [13] Now is the time to recognize the inter–connectedness and complementarity of the religions, as seen through the metaphor of the Ganges, according to Panikkar.

At the outset, we must state our skepticism with Hick and Knitter's third option of "pluralism". At least two things concern us at this point: the nature of the pluralist position itself, and the issue of the "other". Christoph Schwöbel has written,

> The pluralist approach, contrary to its avowed intentions, seems to tend to develop a picture of the universal and ultimate noumenal focus of all religions transcending the particular concrete religions or of a common anthropological constant

> underlying all particular religious expressions, which allows their distinctive particularity only a penultimate and preliminary status. [14]

This statement summarizes the main difficulty with the pluralist position, namely, that of particularity. Pluralism attempts to enjoin together all religious experience into a "numinous" category that is neither as distinct as the religions are nor as postmodern as its epistemology needs to be. Even John Cobb has difficulties with how this definition of pluralism espoused by Hick, Knitter, et al., is achieved:

> Probably the most basic assumption is that there is an essence of religion. This essence is thought to be both a common characteristic of all "religions" and their central or normative feature. Hence, once it is decided that Buddhism, Confucianism, or Christianity is a religion, one knows what it is all about and how it is to be evaluated. The next step is then the one about which the consensus was to be formed. Given the common essence, let us agree to acknowledge that it is realized and expressed more or less equally well in all the great religions. [15]

Cobb then goes on to illustrate his contention by pointing out some basic differences in religions. For example, in the Abrahamic faiths, the worship of a Supreme Being is central to its spirituality. However, in most Buddhist traditions this trait is absent, yet it is still qualifies as a religion because of its deep concern for the character of its adherents. Similarly in Confucianism, "by a certain stretch of terms one can find in it a worship of a Supreme Being, but the function this plays is far less central than in Judaism, Christianity, and Islam. There is great concern for the right ordering of human behavior, but much less interest in transforming the quality and character of experience." [16] Very great differences in the religions remain, and any pluralism that refuses to recognize the particularity inherent in them puts its universality at risk. [17]

For instance John Hick has argued that all religions understand salvation as shifting from self–centeredness to reality–centeredness. [18] And while we cannot know Reality in full, the commonality for religion is seen in *agape* or *karuna*. [19] This sounds plausible, until the Christian concept of *agape* is examined further, and at that point, even a pluralist such as Ninian Smart realizes some problems: "humility may loom large among Christians where the root sin is pride or ego–inflation: but it may not be important for Muslims or Brahmins and indeed may be thought counter–productive in a number of cultures." [20]

In addition, as we have already mentioned, the postmodern viewpoint wields its head at the very moment when pluralism seems to have accounted for the differences in religion. In fact, by deconstructing the religions down to one essential point or a group of points, pluralism is guilty of what it accuses *exclusivism* of doing – failing to take the particularity of each religion seriously, and in so doing pluralism exhibits its thoroughly modern roots. This is the new imperialism of a metanarrative gone amuck, just when we were to be freed from the imperialism and domination of the Christian metanarrative. Gavin d'Costa remarks that "the 'system' of pluralism [seems] to operate in a curiously absolutist fashion, proposing to incorporate religions on the system's own terms rather than on terms keeping with the self–understanding of the religions." [21]

The challenge of the pluralist enterprise, it seems, is to retain the particularity of one's own system, while still respecting the "otherness" of other religions. Kathryn Tanner suggests pluralism does not respect the "other": "In imitation of the general way colonialist discourse constructs its 'Others,' the pluralist insistence on identity in beliefs, norms, or reference as a presupposition for inter–religious dialogue undermines, I argue, respect for other religions *as other*." [22] The "Other" may or may not have a system that is "unique" or "absolute" or "true", nevertheless its particularities must be laid bare in comparison with one's own particularities; or as Kevin Vanhoozer states, we "must not subsume but *submit* to the Other insofar as we must be willing to put our beliefs to critical tests." [23]

Hick himself replies to these critiques and gives his perspective in his *A Christian Theology of Religions* (1995). Addressing the postmodern concern, Hick writes:

> A major theme of some of the post–modern writers is their attack upon all 'totalizing' thinking that would force the complex and variegated world into a single conceptual scheme. Such thinking writes a global meta–narrative that subordinates all individual and communal narratives, thus undermining 'alterity' and eliminating the otherness of the Other. And critics apply this to religious pluralism, which they say 'totalizes' the religious realm, suppressing the mutual othernesses of the religions. Religious pluralism, says one writer, "is a comprehensive and homogenizing historical scheme which assimilates to itself, and thereby tames and domesticates, the practices and beliefs of the different religious traditions"... Indeed, even more fundamentally, doesn't religious pluralism assume that the religions are all trying to answer the same question, whereas in fact they are each asking their own different question? [24]

Hick answers, "Yes, it's true that the religions ask different questions. But I want to suggest that these questions, whilst *specifically* different, are *generically* the same. They all presuppose a profound present lack, and the possibility of a radically better future; and they are all an answer to the question, how to get from one to the other. In traditional Christian language they are all ways of asking, 'What must I do to be saved?'"[25] Hick further explains his position: "the practical outcome is not that there should be a new global religion, the same for everyone, but that the adherents of each of the existing world faiths should respond as fully as possible to the Real, the Ultimate, in their own way by devoutly living out their own tradition." [26] When pressed with the observation that Hick is proposing a "meta–theory" for all religions, Hick agrees, but he is still arguing for the validity of different doctrinal systems within their own religious systems. He escapes this conundrum by positing that the "differing belief systems are beliefs about *different* manifestations of the Real. [27]

We have noted Hick's response in some detail because his scheme, and the critiques of it, are crucial to our point of view in this work. Hick himself notes that "not more than one of these rival belief–systems could be finally and universally true". [28] And this, we believe, is the major flaw in his position – he is somehow attempting to hold to both a universal, overarching Real, and the dizzying plurality inherent in the many religions. By holding to the former, he has implied that the latter is inconsequential. How, in fact, can the all–powerful Allah of Islam be harmonized with the no–God of Buddhism? How can the pure emptiness of nirvana be complementary with the worship of the Lamb in Revelation? How can the many, many gods of Hinduism be meshed with the strong monotheism of Yahweh? Hick's perspective does not answer these types of particularities with any consistent satisfaction. In fact, Hick's own attempt at developing a pluralistic Christian theology of life after death exemplifies some of the main difficulties associated with his position – an overarching scheme that picks and chooses things from differing religious traditions without harmonizing the very pointed differences in eschatologies. [29] Hick has failed to fully appreciate the "other", and in the final analysis, his scheme has foisted an overarching theory of religions that is scarcely better than the colonializing hegemony of modernity. His theory is in fact a vestige of modernity, but one that crumbles in the face of postmodernity.

What is needed then is a theological statement that rises from within a particular tradition (so that it is true to the details of one's tradition) and that is in discourse with other traditions (so that it is not guilty of insulation and not respecting or understanding the "other" in this era of postmodern diversity). This is precisely the point made by Dupuis, who argues that a theology of religions must either be "confessional," in the sense that it is committed to a religious faith, or it does not exist at all. [30] We may extrapolate Dupuis' statement to theology as well: If a theology is not "confessional" – that is, linked to a tradition, it does not exist at all. We will elaborate on this theme as we proceed. The adoption of a particular tradition when doing pluralistic theology does not preclude, however, the second part of our task – discoursing with other traditions. This discourse is not only necessary because of the era in which we live, it is necessitated by the global thrust of Christianity itself. Dupuis describes our aims well:

> What is called for is not a common theological enterprise which levels differences in a search for a common denominator, but the frank admission of the plurality and diversity of beliefs and the mutual acceptance of the others in their very otherness. The model that needs to be developed then is not that of mutual assimilation through a reduction of faith–content but that of interpenetration and cross–fertilization of the various traditions in their diversities; not a leveling of religious identities but a dialogical openness and mutual enrichment through conversation. *Personal commitment to one's own faith and openness to the faith of others need not be mutually exclusive; rather they ought to grow in direct proportion* (italics mine). [31]

This will be our task in this book – to both respect and learn from other traditions, while retaining the distinctives of the Christian perspective. We then can humbly offer our work to the world, confident in the moorings of our own tradition, yet open to learning from the "other" or "others."

With this information as background, we may proceed with some definitions. First, "pluralistic theology" refers to an enterprise in Christian theology that consciously attempts to dialogue with other religious traditions in crafting theology. We will be advocating for the process not only of dialogue, but also, as suggested by Dupuis, the process of "interpenetration" in our pluralistic theology. Secondly, besides referring in a generic way to the entire arena of differing religions, "pluralism" (or "pluralist") refers to a position with respect to how one does pluralistic theology – namely, the third position of Hick and Knitter. This position does not assume any finality to the Christian position, and is in general, rejected by our viewpoint. Our task in this dissertation will be to do pluralistic theology without advocating pluralism as we have defined it. Third, we must mention the difference here between a "theology of religion," and a "theology of religions." A theology of religion "asks what religion is and seeks, in light of Christian faith, to interpret the universal religious experience of humankind." [32] A theology of religions "studies the various religious traditions from the perspective of Christian faith." [33] Our goal will be to build upon some of the understandings of a theology of religions (especially in chapter four), but to move beyond methodological considerations to theological incorporation.

Outline of Work

Rather than the hegemony of the pluralism of Hick and others, or the complete relativism of certain radical postmodern theologians (cf. Mark Taylor, Thomas J. J. Altizer, etc.), we want to offer another model of postmodern pluralistic theology, one that respects and learns from each tradition while still remaining rooted in one's own. This methodology takes dialogue more seriously (in fact, it takes it a step further – towards incorporation), while at the same time taking each religious tradition more seriously and not homogenizing it for the sake of a pluralistic construct. This pluralistic theology will be thus a truly Christian theology, but not one that is in isolation from other viewpoints. And in fact, that is where this book will end, with a constructive Christian statement on the doctrine of creation informed by other religions. But to get to that end, we must develop a methodology for how we can not only dialogue with other traditions, but move towards *incorporating* material from them into a Christian theological doctrine. Our task here will be to develop that model, then put it into practice in the area of creation.

This methodology has been done in a limited way in McDermott's *Can Evangelicals Learn from World Religions* (2001), Tennett's *Christianity at the Religious Roundtable* (2002), and Corduan's *A Tapestry of Faiths* (2002), but what we are advocating for is an even fuller adoption of a pluralistic theology methodology in some of the major areas of systematic theology – God, creation, humankind, etc. – from a Christian viewpoint. This theology should move beyond "encounter" in dialogue with other traditions, to "incorporation." So, to that end, this book will help develop that methodology, as well as use some major themes in creation as an example of that methodology.

The doctrine of creation is a significant theme for systematic theology and for Old Testament theology as well. So a fuller statement on a doctrine of creation is warranted in its own right, given the paucity of full–length treatments on it. But beyond that, the doctrine of creation's roots "in the beginning", for all people, constitute an irresistible springboard from which to do pluralistic theology. What's more, given the incredibly pluralistic context *out of which the Old Testament's own doctrine of creation emerged*, we have an example of how the Christian tradition crafted her viewpoint in the face of competing viewpoints. [34] Furthermore, later in the Old Testament corpus, we find a method for how to do the type of pluralistic theology we are advocating. Thus, we will attempt to show how Israel not only was aware of other (competing) religious traditions, but how she incorporated them into her own theology. After delineating this model, we will then return to the doctrine of creation and put the model to use by incorporating insights from other religions into a pluralistic Christian doctrine of creation.

An appropriate metaphor for this endeavor is that of a journey. This is not a static investigation of a few concepts or a few authors, but rather an experimental probe into the development of a methodology that will lead us into a wide variety of areas in theology – epistemological, systematic, cultural, historical, biblical, and constructive. These six headings form the rough contours of our work. We will begin with the epistemological and systematic warrants for our enterprise – postmodernity and the arena of natural theology, respectively. These areas will show us how we can proceed with our inquiry. Once we have outlined an epistemological and theological rationale, we will explore two very different current doctrines of creation – Colin Gunton's and Ninian Smart's – as a way of explicating the effects of being open to the viewpoints of other religions. Gunton is exemplary of a doctrine of creation that is a self–contained Christian statement, while Smart's is an avowedly pluralistic doctrine. Taken together, Gunton and Smart, along with other religions' cosmogonies constitute what could be seen as "cultural" theology. We will then visit the Christian traditions' history of using material from other religions in an exercise in historical theology. Finding a paucity of models there, we will examine a model from Scripture itself – Old Testament Wisdom Literature – for integrating doctrines from other faiths. This biblical theology will give way to a delineation of key areas in other religions' cosmogonies, and an articulation of possible areas for Christian re–examination. Finally, we will move to our constructive theology, where we will use our model and move forward to pursue three crucial areas of a

Christian doctrine of creation that are informed by the worldwide religious community.

Such a constructive pluralistic theology of creation is largely unique in moving beyond the theoretical possibility of incorporating insight from other religions to actually making use of the insight of other religions in formulating Christian doctrine. We will discover that while many theologians have a theology of religions that is open in principle to material from other religions, they rarely take the next step to adapt or even examine any such material for their own Christian scheme. Curiously, two of our main examples are located in studies of eschatology – one would be John Hick's *Death and Eternal Life* (1976), from a more avowedly pluralistic viewpoint; another would be S. Mark Heim's *The Depth of the Riches* (2001), from a more inclusivistic viewpoint. But in the main, there are precious few theologians who have interacted with material from other religions on a major systematic theology topic. Most stay safely in the realm of dialogue, without ever venturing to incorporate any insights from other religions into their theologies. So rather than capitulating to the lure of pluralism, or on the other end of the spectrum, refusing to examine what many profess to be an example of God's general revelation, we are attempting to do justice both to our own tradition and God's working in the wider world by putting forth a truly pluralistic Christian doctrine of creation.[35]

CHAPTER 2

Rationales

Epistemological: Holism

Postmodernity is not only our context but also contains our epistemological rationale for moving in the direction we are suggesting. In this chapter we will be examining the possibility of holism, as opposed to foundationalism, as our operating epistemology in this new postmodern world. While some theologians suggest postmodernity is simply one of the latter phases of modernity (cf. Jean–François Lyotard [1]), our position is that postmodernity, as an epistemological era, is distinct. It has arrived and is operative. What is this postmodernity? For a definition, Robert Pippin suggests we look at Arnold Toynbee's account of the fourth and last phase of history, designated "postmodern", which involves irrationalism, anxiety, and lost hope. [2] Pippin himself offers an underlying sentiment: "Whereas in modernism, the typical modern experience that 'all that is solid melts into air,' or 'the center does not hold,' had prompted the creation of a 'subjective center', an autonomous self–defining artist, for postmodernism there is no center at all." [3] Gregory Smith goes on to augment this definition:

> Contemporary postmodernism is fundamentally a sign of disintegration, of transition, of waning faith in the modern ideas of Reason and Progress, and the Enlightenment project in general. It lacks faith in the modern autonomous subject as self–grounding and self–legislating. [4]

In light of these definitions, postmodernism has been seen fairly negatively from some points of view. This disintegration that Smith speaks of is seen as a great challenge. Os Guiness is perhaps representative of this skepticism when he writes, "The trademark gesture of postmodernism is not a handshake, a hug, or a punch. They are each too committed. It is a wink and a smirk." [5] This negative characterization is emblematic of the sense that seems prominent among some of keeping postmodernity and its "forces" at bay. Our approach in this dissertation will be one that assumes the "take–over" of modernity by postmodernity, and just as theologians have had to deal with the consequences of modernity, so too must we all negotiate with postmodernity. Interestingly, the demise of modernity may mean the demise of theology built on modernity, so that postmodernity represents a chance to rebuild theology that was perhaps too bound up with modernity, and even to retrieve much of pre–modern theology that modernity wrongly threw out. So, far from being interested in postmodernity from a "faddish" point of view, or seeing it as primarily a negative force in theology, postmodernity offers an escape hatch, if you will, from the confines, strictures, and perhaps even the heresies of modernity. [6]

One of the constitutive elements of postmodernity is its epistemological approach. Foundationalism was a key tenet of modernity's epistemologically, but we must develop our own postmodern epistemology that we can put to use in developing a pluralistic Christian theology. We will use the imagery of holism for our approach, but before we explicate that, we must examine how foundationalism developed, and its weaknesses that perhaps a new epistemology can correct.

Modernity and Foundationalism

Modernity has been said to begin somewhere between the Renaissance and the Enlightenment, with a central element being the elevation of humanity to the center of existence. Philosophically, the period is said to begin with Rene Descartes, who is often thought of as the first modern philosopher. His death in 1650, however, precedes by 150 years the supposed beginning of modern theology, which is often dated from the publication of Friedreich Schleiermacher's *On Religion: Speeches to Its Cultured Despisers* in 1799. [7] For our purposes, we need to center on one of the crucial mooring points of the modern mind, foundationalism.

Let us return to Descartes who is so representative of the foundationalism of modernity. Descartes set out to re–examine or doubt all of the beliefs he had been taught, and thus build his knowledge on indubitable principles. [8]

In terms of a theory about knowledge, "foundationalism", as it has come to be known, works on *basing* beliefs on certain unassailable principles. Nancey Murphy writes of foundationalism:

> When we seek to justify a belief, we do so by relating it to (basing it upon, deriving it from) other beliefs. If these other beliefs are called into question, then they too must be justified. Foundationalists insist that this chain of justifications must stop somewhere; it must not be circular or constitute an infinite regress. Thus, the regress must end in a "foundation" of beliefs that cannot themselves be called into question.[9]

Theologians have recognized increasingly the role that this foundationalism has played in the development of modern theology. Murphy herself believes that foundationalism has forced theology to build itself on two main cornerstones – either Scripture for conservatives, or experience for liberals. [10] An example of this is seen in propositional theories of scripture, in which propositions themselves form the basis from which all other claims of religious knowledge are derived. For example, Carl F.H. Henry writes, "Scriptures contain a body of divinely given information actually expressed or capable of being expressed in propositions." [11] These propositions then form the indubitable foundation upon which revelation, truth, and humanity's response rest. A little further to the left of this position is the foundationalism of Donald Bloesch: "As evangelical Christians we can and must speak of foundations of the faith. These are not, however, a priori principles or self–evident truths but the mighty deeds of God in the history of biblical Israel . . . ." [12] For Bloesch the propositional nature of the texts isn't the foundation, as it is for Henry, but rather the foundational aspect of scripture is God acting in history.

On the more liberal end of the spectrum, beginning with Schleiermacher, we have the dominance of human experience as a foundation. In *The Christian Faith*, Schleiermacher argues for faith as a "feeling of absolute dependence", and refuses to speak about God apart from our human experience of him. [13] Thus human experience exemplified either through reason or emotions provides the basis for all subsequent religious thinking. David Tracy provides a more contemporary example of this type of foundationalism in his *The Analogical Imagination*, where he writes of the Christian scriptures as literary classics. For the reader, the classic is an expression of *experience*: "It remains the realized experience which must serve as the first word and the final criterion of relative adequacy in any attempt at both understanding and explanation."[14] Schleiermacher and Tracy's experiential foundationalisms are merely liberal expressions of a theory of belief justification rooted in modernity, just as Henry's and Bloesch's are more conservative expressions of this same belief structure.

The Inadequacy of Foundationalism

If foundationalism has been so prevalent, why then is it now being called into question? It is not adequate to say postmodernity has rendered it moot, rather we must delve into why foundationalism has been seen to be inadequate as a theory of belief justification.

Part of what the concept of foundationalism is getting at is the idea of a universal truth, applicable to everyone. Indeed the imagery of a foundation assumes one building for everyone. Stephen Toulmin hints at this quest for universality as he speculates on the reasons why Descartes wrote as he did. Toulmin notes that Descartes wrote after the Thirty Years War (1618–1648), and with all the bloodshed and religious differences present, Descartes was looking for a path of universal agreement. If human reason could provide a roadmap for universality, then the direction after Descartes could focus on the universal, the general truth, for people everywhere to assume.[15]

This aspect of foundationalism, namely, universality, has come under serious fire in the eyes of postmodernity. What modernity promised in terms of universality of thought has run headlong into the particulars of a world where not everyone agrees with indubitable principles, much less Descartes' ideas of universal principles. The universal has become the particular, the local. Reason itself has been called into question and language, once securely referential, has been critiqued as well. Through critical theorists like Foucault and Derrida, the very notion of ideas representing something real has been dealt a heavy blow. So that to claim to represent universal ideas in our age is to spar with a very prevalent and palpable sense of pluralism.[16] Indeed, Graham Ward writes of postmodernity,

> Words like development, progression, advancement, meaning, profundity, and depths are supplanted by other words like dissemination, indeterminacy, deferral, aporia, seduction, and surface. Meaning is local, community is tribal, society is pluralistic, and economics is the pragmatics of the marketplace.[17]

A second major reason for abandoning foundationalism as an epistemological system has been the recognition of the interpreter's role in knowing.

And this is related to our first difficulty cited, namely modernity's assumption that universal truth is discoverable via a neutral method. Hans Georg Gadamer has been instrumental in showing how it is nearly impossible to start with a "clean slate", and that the modern attempt to define truth apart from particularity had to fail. Reacting against the idea that total objectivity was required in hermeneutics and following Heidegger, his teacher, Gadamer developed his idea that understanding is like a dialogue where we use our prejudices and commitments – he calls them "fore–understandings" or "fore–concepts" – in conversation with the actual texts. [18] In other words, neutrality is unachievable, especially in the case of sacred texts, where Gadamer argues that the hermeneutical process is made up of three things: 1) the horizon of meaning surrounding the practical questions; 2) the fore–understandings we bring; and 3) the horizon of meaning the texts themselves project. [19] The consequence of this perspective is that if we ourselves are involved in every interpretive and creative process, the only foundation that can be laid is ourselves. And someone else will undoubtedly lay a foundation unique to his or her person, so that universality vanishes and all that is left is the particular, or the individual.

Germane to this point is the idea of "theory–ladenness" that Thomas Kuhn and others have discussed. "Facts" themselves are dependent not only on interpretation, but on theories built into the apparatus for discovering those "facts." For instance, an electron microscope that produces images is itself the product of a previous theory. [20] So a claim for universal *principles*, in light of this critique, is fraught with the danger of confusing the individual with the general, and the abstract and "pure" with the assumed and dependent.

In summary, let us simply reiterate what Jeffrey Stout has written: "The era of foundationalist epistemology, the era in which epistemology was the basic concern of philosophy, is now as a matter of historical fact over." [21] This is cause for neither celebration nor concern, but for opportunity, as we shall see. But it has been necessary to establish the context into which a truly postmodern, nonfoundational understanding of theology must emerge. [22]

An Alternative to Foundationalism: Holism

Before we leave our discussion of postmodern epistemology, let us offer a model of nonfoundationalist epistemology. W.V.O Quine has rejected a foundationalist epistemology for a holist one in his crucial essay, "Two

Dogmas of Empiricism".[23] In this theory, knowledge is less like a foundation and more like a web or a net. In some interesting parallels with Gadamer, Quine argues that what we call knowledge is constantly being revised by our experience:

> Total science is like a field of force whose boundary conditions are experience. A conflict with experience at the periphery occasions re–adjustments in the interior of the field. Truth values have to be redistributed over some of our statements. Re–evaluation of some statements entails re–evaluation of others, because of their logical interconnections – the logical laws being in turn simply certain further statements of the system . . . . But the total field is so underdetermined by its boundary conditions, experience, that there is much latitude of choice as to what statements to re–evaluate in the light of any single contrary experience. No particular experiences are linked with any particular statements in the interior of the field, except indirectly through considerations of equilibrium affecting the field as a whole.[24]

This theory of knowledge seems to make better sense of how we all come to "knowledge": We experience something; it fits or it doesn't fit with our previous theories about the world; we revise our theories or reinterpret our theories in light of our experience; our theories themselves are a product of our communities of reference, etc. This is a *web* of knowledge, rather than a foundation. Nancey Murphy has offered three comments on this theory of knowledge:

1. Holism refers to the linkage and support of beliefs with other beliefs, and ultimately to a whole belief system.
2. Holism does not need indubitable foundations. Indeed, there are not any "sharp distinctions among types of belief, only degrees of differences in how far a belief is from the experiential boundary."
3. The reasoning in holism does not only go up as it does in foundationalism. Rather, reasoning can go forward or backward or sideways depending on logic, conclusions, and presuppositions. "This allows holists to take account of the fact . . . that the data (scientific facts, interpretations of texts, or whatever) are theory–laden, partially dependent on theoretical knowledge."[25]

This type of epistemology provides the backbone for our exploration of other sources outside the Christian tradition. If a web is a better epistemological model, then it makes sense that one would consider a wide variety of sources of knowledge rather than just one (i.e. foundationalism). This would seem to open one up to the danger of not being true to any tradition, which is the bane of contemporary pluralism. However, our proposal suggests that the web has a center, a core – in this case, the Christian tradition – that is the source from which to explore knowledge.

One of the challenges with this type of theory is to avoid epistemic relativism. If all we have are a series of webs, how do we decide which web is better than the others? There is no basis for comparing the two, it is said. And this is a great difficulty of postmodernity for many – that the pluralism latent in a nonfoundationalist epistemology leads to relativism. [26] While we are not able to fully discuss this claim here, we must mention the work of Alasdair MacIntyre, whose account of rival traditions of moral enquiry is in agreement with much of Quine's philosophical reflection on holism in MacIntyre's important work, *Three Rival Versions of Moral Enquiry*. MacIntyre has argued that it is indeed possible to compare traditions and rationality, based on the *traditions themselves*. In other words, a tradition (or web of beliefs) is justified if it succeeds or fails within the tradition's own internal standards of rationality. [27] For example, within the Christian community, MacIntyre claims Aquinas' synthesis of Augustinianism and Aristotelianism in the thirteenth century showed the superiority of the Augustinian tradition since Augustinianism was able to adapt, to synthesize, while still remaining true to its basic tenets. Aquinas' rivals failed, according to MacIntyre, because their epistemology and rationality were inferior to Thomas' Augustinianism. In other words, each tradition or web has internal criteria for judging its truth or reliability.[28]

MacIntyre's work is key for us, in that we see in a postmodern, post–foundationalist setting, the crucial role a tradition's own consistency and coherence plays. This is the center of our web of knowledge; moving out beyond that center, in a holistic epistemology, we encounter other traditions, not only in our own religion – Christianity – but in other religions as well. Thus postmodernity helps to not only shape our method of theological inquiry, but also to strengthen the role of the tradition that is at the core of our exploration. Rather than offering only skepticism and *ennui*, postmodernity, and in particular, postfoundationalism, affords us an opportunity to both reshape and present our views to the wider community, and in a roundabout way, connects with how we will do pluralistic theology in this book.

To stand, then, confidently (but not arrogantly) in the Christian tradition, with an eye toward the wider world, signifies not only our stance on pluralistic theology, but also our theological task in these postmodern times. Elizabeth Barnes argues that one can make universal claims, but that that process is far more effective from within the framework of one's own narrative, or as Nancey Murphy comments, if the "epistemological warrant is particular." [29] Similarly Terrance Tilley suggests that our method should be that of mutually seeking truth, but that that search should not be from the position of an uncommitted, neutral academic. Since practice requires no external epistemic foundation, we should engage in the "unconstrained communication" (cf. Habermas) and "conversation" (cf. Rorty) of dialogue, but not under the pretense of noncommittment. [30] Trevor Hart describes what this postmodern theological task would ultimately look like:

> The Christian theologian, like Polanyi's scientist, confesses the partial and commitment–laden nature of the knowledge which he offers to others. He makes no unrealistic or self–deceiving claims concerning its certainty or universal demonstrability. Yet, as one who genuinely believes himself to have encountered, or rather been encountered by, the supremely 'public' reality of the God who is Lord of all, he finds himself placed under a moral obligation to speak of this reality, to commend him to others, and to do so by commending to them the framework of the Christian story as the particular location where this God has made and makes himself known. His commendation of this particular tradition, this framework of meaning, is not intended to deny in any arrogant manner that this reality might be known in other ways and other places... At the end of the day, therefore, the reception of this knowledge by others as public truth will depend upon their willingness to entertain this same standpoint, to consider this particular story and seek to view things from within it, to explore the heuristic potential which is claimed for it. [31]

This, we suggest, is what postmodernity offers theology, a chance for a theology that is not only true to its own tradition, but one that is in genuine dialogue with others.

Theological: Natural Theology

If postmodernity paves the way for us epistemologically, and if we are to examine material in other religions so as to include them in a Christian theology, what is our theological rationale for doing so? Contemporary pluralism seems to leave aside the heavy theological lifting involved in developing such a rationale, and moves quickly to theologies of religion and questions of truth, representation, dialogue, etc. The basic question is typically not asked, namely, can the Christian God be involved in, or speak through, other religions? And if so, how? The concern of method is not addressed, as it was in earlier generations of theologians who spoke of natural theology and general revelation. We would like not to ignore such a crucial step, but instead go back and develop a center, a method, from which our theology can proceed, and so we must address these issues of natural theology and general revelation, because these are really the two main areas that establish theologically our right to proceed in this direction.

To do that we will go back into church history and revisit three seminal eras – scholasticism, the reformation, and neo–orthodoxy. Our goal will be to see how these issues have developed, and where past theologians have guided the debate in these areas. We hope to provide strong theological criteria for moving in the route we are headed.

Thomas Aquinas

While Augustinian thought that had its roots in Platonism came to be a dominant thought form in the Western church, it was up to Thomas Aquinas to develop and adapt that other great philosopher – Aristotle – into a synthesis of faith and reason known as scholasticism. This area of "reason" is important for us, because in it Aquinas develops some of his most important work on "natural theology." Colin Brown has defined natural theology as "the attempt to attain an understanding of God and his relationship with the universe by means of rational reflection, without appealing to special revelation." [32] Brown's definition highlights for us the relation of natural theology to general revelation. General revelation can be defined as "God's self–disclosure and self–communication in the universe and created world." [33] Natural theology is thus a theology built upon the implications of a robust theology of general revelation: that God is working throughout the entire creation. Brown's definition also suggests a core issue with natural theology, namely, its attempt to understand God – can God be known (and thus worshipped, followed, or satisfied) through general revelation alone, or must we resort to special revelation to truly know God? This relates of course to a central axiom with pluralism, and that is its avowed step of assuming that God is working in other religions. Our contention is that these areas – general and special revelation – and the symbiotic relationship between these areas, need to be developed further in order to truly ground our endeavor in the Christian theological tradition, and thus further strengthen our holistic web of knowledge.

Regarding the importance of reason, Aquinas' work stops short of a fully developed "natural theology," that is, a theology which contends that it is possible to understand fully and relate to God outside of specifically Christian revelation. Aquinas believes that reason can be used in the ascertaining of some truths of the Christian faith: "Such are that God exists, that He is one, and the like. In fact, such truths about God have been proved demonstratively by the philosophers, guided by the light of the natural reason." [34] Aquinas does not believe human reason is without limitations, however: Human reason is inadequate and divine revelation is needed because of 1) the depth and subtlety of the object [God], who is far removed from the senses; 2) the weakness of human intellect when it begins its operations; 3) the length of time required to learn things needed for a conclusive proof. [35] Faith provides a greater certainty than reason, and love and grace are necessary to perfect faith: "Thus we must say that for the knowledge of any truth, man needs divine assistance so that his intellect may be moved by God to actualize itself. He does not however, need a new light, supplementing his natural light, in order to know truth in all cases, but only in certain cases which transcend natural knowledge. . . . The human intellect cannot know more profound intelligible realities unless it is perfected by a stronger light, say the light of faith or prophecy; and this is called the light of grace, in as much as it supplements nature." [36] Faith then is "beyond reason," in Aquinas' famous phrase, but it is not opposed to reason: "although the truth of the Christian faith which we have discussed surpasses the capacity of the reason, nevertheless that truth that the human reason is naturally endowed to know cannot be opposed to the truth of the Christian faith." [37] Arvin Vos summarizes Aquinas well:

> We have been trying to show both that Aquinas has a place for natural theology and that he keeps it strictly subordinate to revealed theology. He holds that what the philosophers can grasp is properly speaking a preamble to faith – and by a preamble he does not mean something that is necessarily or even normally prior to the knowledge of faith; indeed, he argues that it is far more typical that one will *believe* that God exists before coming to *know* that he exists .... By working through to the conclusions of natural theology, believers can supplement their knowledge of things grasped by faith without any risk of making revealed theology dependent on or subordinate to natural theology. [38]

What we see then is a carefully nuanced picture of the relationship between faith and reason. In fact, faith and reason are to be complementary. There is a place for reason, or natural theology, for Aquinas; however it is strictly subordinate to revealed theology, or faith. Natural theology, or such that the philosophers can grasp by reason alone, is, properly speaking, a preamble to faith. [39] Thus for Aquinas, grace perfects nature, and faith perfects reason. It fell to later interpreters of Aquinas to over–emphasize his work on reason such that for many Protestants, Aquinas' work became synonymous with a full–fledged natural theology. [40]

But for our purposes, what Aquinas has done is set the stage for the possibility of learning about God without an appeal to (special) revelation. Aquinas' emphasis on the role and possibilities of reason would provide an enduring methodology for theologians attempting to not only discuss and expound on general revelation, but also to provide the basis for dialogue with other religions. [41]

John Calvin

As we move through church history, we come to the figure of John Calvin, the 16th century reformer. Both Luther and Calvin were consumed with explicating the Christian faith in a time of great dissatisfaction with the Roman Catholic church, resulting in what we now call Protestantism. Calvin comments briefly on the place of other religions in Christian theology, even though his high Christology prevents him from acknowledging truth in their worldviews. He also offers some tantalizing insights on the issue of natural theology, though he certainly does not develop his thoughts in this area.

But it is in Calvin's thoughts on the nature of revelation that we see his clearest picture as it relates to our topic. In the beautiful fabric of the world, and the movement and order of the heavenly bodies, "the glory of God shines forth more clearly". [42] Indeed, the image of Calvin bookishly hiding away from the joys of nature does not square with his effusive praise of creation:

> How great must have been the Artist who disposed that multitude of stars which adorn the heaven in such a regular order, that it is impossible to imagine anything more beautiful to behold; who fixed some in their stations so that they cannot be moved; who granted to others a freer course, but so that they never travel beyond their appointed limits; who so regulates the movements of all that they measure days and nights, months, years, and seasons of the year. [43]

Even further, Calvin suggests, "He hath so manifested Himself in the whole workmanship of the world, and daily shows Himself openly, that men cannot open their eyes without being forced to see Him." [44] This manifestation,

Parker concludes, represents a "single well–defined idea in Calvin's theology. The universe is a mirror in which is to be seen the *effigies Dei*, the portrait of God … Thus, in creating the universe God made it a representation of Himself." [45] Not only does nature display God, but "Calvin admits that the rays of Christ are diffused throughout the whole creation," writes Wallace. "[Calvin] even suggests that pagan culture and religion owe much to Christ." [46] Calvin also acknowledges that there is a kind of natural knowledge of God in humanity: "There is within the human mind, and indeed by natural instinct, an awareness of divinity." [47] So that we may say, by way of summary, that the knowledge of God is *potentially* readily available to all of humanity in a variety of ways.

However the possibility of this knowledge alone cannot lead us to God.

> Vain, therefore, is the light afforded us in the formation of the world to illustrate the glory of its Author, which, though its rays be diffused all around us, is insufficient to conduct us into the right way. Some sparks, indeed, are kindled, but smothered before they have emitted any great degree of light. Wherefore the Apostle, in the place before cited, says, "By faith we understand that the worlds were framed by the Word of God"; thus intimating, that the invisible God was represented by such visible things, yet that we have no eyes to discern Him, unless they be illuminated through faith by an internal revelation of God. [48]

There is needed a special revelation of God – Christ – to help us know God. Thus our own efforts, reason, and industry are not enough to lead one to Christ. Wallace concludes:

> 'Even philosophers,' says Calvin, 'contributed nothing whatever that they might glorify God. All they say regarding religion is frigid and insipid. It is therefore in His word alone that there shines forth the truth which may lead us to true piety.' Calvin had a high respect for the liberal arts and sciences but the cultivation of these cannot lead us towards God. 'Natural reason will never direct men to Christ; and as to their being endued with prudence for the regulation of their lives or born to cultivate the liberal arts and sciences, all this passes away without yielding any advantage.' Calvin admires in Daniel his superior natural abilities and learning, and his endowment with remarkable intellectual gifts, but points out that even then it was not through his own human industry or exertion that he acquired his spiritual knowledge. That knowledge came through prayer. [49]

In the end Calvin believes that reason, nature, or exertion will not lead one to a true knowledge of God.

Even the inherent and indelible *sensus divinitatis* in humanity, which Calvin readily admits to, cannot lead one to God:

> While experience testifies that the seeds of religion are sown by God in every heart, we scarcely find one man in a hundred who cherishes what he has received, and not on in whom they grow to maturity, much less bear fruit in due season. Some perhaps grow vain in their own superstitions, while others revolt from God with intentional wickedness; but all degenerate from the true knowledge of him. The fact is, that no genuine piety remains in the world. [50]

This knowledge is extinguished or corrupted, partly by ignorance, and partly by wickedness. This theological hurdle – the role of sin – is a key critique of natural theology that will crop up in other explications of natural theology, and we will return to it later. The key here is that the sinfulness of humanity has rendered it unable to find God: "At length [people] involve themselves in such a vast accumulation of errors, that those sparks which enable them to discover the glory of God are smothered, and at last extinguished by the criminal darkness of iniquity." [51] Only divine revelation, located principally in Christ, can lead us to God.

We can summarize that for Calvin, though there is the potential for learning about God from sources outside the Christian tradition, that potential is never really actualized, or if it is, it is of miniscule value. Thus Calvin does not hold a place for natural theology, though perhaps he articulates a theology of nature. [52]

Calvin, however, may have underemphasized the role of philosophy and "natural learning" in his own teaching and ministry. William Bouwsma writes, "between 1527 and 1534, at any rate, and in a more general sense all his life, Calvin inhabited the Erasmian world of thought and breathed its spiritual atmosphere; he remained in major ways always a humanist of the late Renaissance." [53] Bouwsma notes that Latin oratory (i.e. Cicero and Quintilian) was the model for would–be pastors, the great humanist Erasmus' educational program was used in pastoral education at Geneva, and even the use of the word *Insitutio* was representative of a humanist of that era. [54] However, this influence, though great in the foundations of Calvin's epistemology or in the organizing of his catechism, is met head on by his thoughts on the centrality of Christ for Christian revelation outlined above. [55]

Karl Barth & Emil Brunner

In the twentieth century we encounter neo–orthodoxy, and its attempt to undermine the reach of nineteenth century German liberalism. Karl Barth and Emil Brunner loomed large in these discussions, particularly in the area of natural theology. There is some usefulness for us in examining Barth's concept of revelation, and the contrasting position of Emil Brunner, Barth's contemporary, made famous in their debate over natural theology. Brunner argues, "wherever God does anything, he leaves the imprint", not only in the world, but also on humanity's conscience, which leads him to state:

> The difficult question is therefore not whether there are two kinds of revelation. The reply to this question must on the basis of Scripture once and for all be a positive one. The question is rather how the two revelations, that in creation and that in Jesus Christ, are related. The first answer – again one which is universally Christian and also obviously biblical – is that for us sinful men, the first, the revelation in creation, is not sufficient in order to know God in such a way that this knowledge brings salvation. [56]

Thus Brunner argues for a natural revelation, but one that does not lead to God in a salvific way. Brunner then goes on to discuss God's preserving grace as part of his revelation to the world, and that God's point of contact with humanity is the *imago Dei*. Contrasting himself vigorously with Barth, Brunner remarks, "No one who agrees that only human subjects but not stocks and stones can receive the Word of God and the Holy Spirit can deny that there is such a thing as a point of contact for the divine grace of redemption. This point of contact is the formal *imago Dei*, which not even the sinner has lost." [57] Brunner believes Calvin and Luther both support this viewpoint of the *imago Dei*, and that it can lead to a natural theology. [58]

Barth's famous reply, "Nein!", contains his perspective on natural theology. Barth defines natural theology as "every (positive *or* negative) *formulation of a system* which claims to be theological, i.e. to interpret divine revelation, whose subject, however, differs fundamentally from the revelation in Jesus Christ and whose method therefore differs equally from the exposition of Holy Scripture." [59] Barth's concern is to adhere to the principles *sola scriptura* and *sola gratia*, and as such he can find nothing evangelical in natural theology: "But apart from that, Brunner would have to be opposed on the following grounds: what he has to prove is that in these passages the life of man in Christ through the Holy Spirit is said to *presuppose* a knowledge of and respect for the true God and that this presupposition forms its *point of contact* in man. Where in Galatians ii and I Corinthians ii has Brunner found anything of the sort? Which of the sixteen verses of I Corinthians ii could be quoted to show that St. Paul wanted to maintain and proclaim another knowledge of God before and beside 'Jesus Christ and him crucified'?" [60]

Barth then enjoins Brunner in his debate over Calvin, arguing that Calvin did not in fact advocate a natural theology. He gives three arguments:

1. "It is true that Calvin spoke of a *Duplex cognitio Domini,* from creation and in Christ .... But in contrast to Brunner he said about a natural knowledge of God through creation only what is said about it in Romans 1:19f.; 2:14f; Acts 14:15f.; 17:24f. He did not regard it as a capacity which man has retained and which has to be reconstituted by faith, as a point of contact for revelation and for the new life in Christ;
2. The possibility of a real knowledge by natural man of the true God, derived from creation, is, according to Calvin, a possibility in principle, but not in fact, not a possibility to be realized by us";
3. "'The knowledge of God which now remains to man is nothing other than the terrible source of all idolatry and superstition' (Comm. in John 3:6; *C.R.* xlvii, 57). It is quite impossible to see how this possibility could in Calvin's theology gain the status and significance of a 'point of contact.'" [61]

Barth believes finally and totally in the inability of humanity to find God on his own:

> But this theological problem is put to us by the truth, presented in the Scriptures, Creeds and Confessions, that man is of himself unable to find access to the revelation of God... How can that 'negative point' be the 'point of contact' for divine revelation, since its demonstration leads necessarily to a triumph of reason at its most natural and unregenerate?... The Holy Ghost, who proceeds from the Father and the Son and is therefore revealed and believed to be God, does not stand in need of any point of contact but that which he himself creates. Only retrospectively is it possible to reflect on the way in which he 'makes contact' with man, and this retrospect will ever be a retrospect upon a *miracle.* [62]

Barth's strong "no" precludes any nuances in any type of natural theology, while Brunner is attempting to put forth a subtle and balanced view of natural theology. [63] Brunner is continuing in the tradition of Aquinas in allowing for a viewpoint of accessing God's revelation in natural humanity, but one that does not lead to a salvific knowledge of God.

A Restatement of Natural Theology

Perhaps what is needed at this juncture in our enquiry is a restatement of a natural theology. It seems apparent that, historically speaking, speculation about natural theology has dwindled since the Barth/Brunner debates of the last century. It also seems clear that many pluralists assume a version of natural theology in espousing their positions, and that that assumption is not matched by any development of the doctrine. From our point of view, a reassessment and reapplication of the doctrine is crucial to establishing our argument in this dissertation. We must adopt a theological method that is squarely in the realm of Christian theology, not assume a method from outside our tradition. Nor should we take uncritically an open stance toward other religions, and find ourselves unwittingly in what Barth called the "history of religions" approach. There is indeed much to be learned from other religions, not solely because of the tenets of pluralism, or because of postmodernity, or because of the increasing diversity of our times. More significantly, our own theology demands it. We cannot skip this step in outlining the resources for Christian theology available in other religions.

So what, then, are the major points to emphasize in a re–visited natural theology? First, we must make some brief mention of the history of the doctrine. It is clear that for most theologians throughout church history, *theologia naturalis* was part and parcel of explaining how God works in the world. Indeed, as Wolfhart Pannenberg has noted, "up to the early part of the 20th century in Protestant theology no one ever disputed either the fact that we have here a different form of knowledge of God from that of the historical revelation in Christ, or the referring of the Christian message to this knowledge by claiming it as a provisional knowledge of the one God whom the Christian message proclaims." [64] Pannenberg's contention includes for him both Luther and Calvin, as well as Melanchthon. In terms of Catholic theology, natural theology has been affirmed since at least Aquinas and was given voice at the First Vatican Council: "The same Holy Mother Church holds and teaches that God, the beginning and end of all things, can be known with certitude by the natural light of human reason from created things." [65] The Catholic Church has thus upheld, more or less, the Thomist position. The first critiques of natural theology as such appear around 1900 in the works of Ritschl and Herrmann, who, following in the wake of Kant's criticisms of the proofs of God, attempt to relegate everything "natural" as irrelevant religiously. [66] Barth, as Herrmann's student, continued in his opposition to natural theology, eventually settling on a position Pannenberg describes in this manner: "Against the synergistic idea of a natural and revealed knowledge of God that harmoniously complement one another, Barth now raised the charge that there was here an attempt to erase the distinction between a theology of the Word of God and a theology based on anthropology. Natural theology was an expression of human self–justification." [67] Consequently Barth uncompromisingly rejects all natural theology.

Is Barth the clarion call back to orthodoxy or can we let stand a more nuanced (and perhaps traditional) view of natural theology? In Pannenberg's view, Barth is the exception in the road to a generalized acceptance of a complementary form of natural and revealed knowledge. We concur. Our position is that natural theology should not be jettisoned *in toto*, but in fact must be retained in a clearly articulated and circumscribed manner as a lynchpin for developing a postmodern, holistic Christian theology. [68] Barth went too far in his critique, and placed himself on the margins of the Christian tradition. We need to re–state this doctrine for our time, and have it be a symbiosis of natural and revealed knowledge. What are the central elements to be retained?

1. The Christian God is revealed to all people through creation.

This is not (or should not be) a controversial statement. It is part of scripture (Rom. 2:14ff), and part of tradition (Aquinas, Calvin, etc.). It does, however, need to be reaffirmed for our time in light of our pluralistic context. The word "creation" refers to the entire created world. By discussing this as general revelation, we must concur with Pannenberg who notes that this sentiment "is not a statement of natural theology. It is a statement that is made about us in the light of the revelation of God in Jesus Christ." [69] As such this knowledge is a facet of the gospel, and not an addendum to "true, revealed theology". By locating this doctrine centrally in general revelation, we can pursue its implications irrespective of Barthian critiques, as well as unlocking its implications for us as we examine other religions.

2. This general knowledge of God is available to all people, but does not lead to God in a salvific manner.

This is what Aquinas, Calvin, Brunner and most Christian theology in the West has stated. Complete and final knowledge of God can only come through "revealed" theology, i.e. Jesus Christ. However we must observe the "level" of revelation present in general revelation. For Calvin, it is miniscule, and is quickly dissipated in the harsh glare of human sin. Brunner seems to give this knowledge more potential by locating it within the *imago dei*. Our sense is that if we believe the complete dissolution of this knowledge is a *fait accompli* in the face of human depravity, then we might as well not even suggest a doctrine of general revelation. It is a non–sequiter. Some helpful knowledge of God must be retained or the doctrine is meaningless.

3. The so–called "point of contact" between humanity and God is not located within the capacity of the human, but rather in the Holy Spirit.

For Brunner, the "point of contact" was in the *imago dei*, which Barth rightly critiques. By placing this capacity within humanity, as a strategic move, this doctrine is then opened up to the strong, and perhaps fatal, blows of the nature of evil and sin. If humanity is fallen, the question of the "fallenness" of the *imago dei* arises. And if the *imago dei* is tainted with sin, as a point of contact with God, can it ever be trusted? [70] Rather than delving into the deep waters surrounding the concept of the *imago dei*, we must ask whether the "point of contact" imagery is the correct one in delineating this doctrine. The "point of contact" epistemology will be continually flawed as long as the human is on one end of the conundrum. It is better, it seems, to locate the so–called point of contact in a theological arena – the Holy Spirit – rather than an anthropological one. This arena has not been explored by most of the theologians writing on natural theology. Calvin, while rejecting (human) reason as a route to divine revelation, does speak to the crucial role of the Spirit of God:

> Whenever, therefore, we meet with heathen writers, let us learn from that light of truth which is admirably displayed in their works, that the human mind, fallen as it is, and corrupted from its integrity, is yet invested and adorned by God with excellent talents. If we believe that the Spirit of God is the only fountain of truth, we shall neither reject nor despise the truth itself, wherever it shall appear, unless we wish to insult the Spirit of God; for the gifts of the Spirit cannot be undervalued without offering contempt and reproach to the Spirit himself. Now, shall we deny the light of truth to the ancient lawyers, who have delivered such just principles of civil order and polity? Shall we say that the philosophers were blind in their exquisite contemplation and in their scientific description of nature? ... What shall we say of all the mathematics? Shall we esteem them the delirious ravings of madmen? On the contrary, we shall not be able even to read the writings of the ancients on these subjects without great admiration; we shall admire them, because we shall be constrained to acknowledge them to be truly excellent. And shall we esteem anything laudable or excellent, which we do not recognize as proceeding from God? [71]

4. The crucial theological core in a "natural theology" is therefore the third person of the Trinity, the Holy Spirit.

Building on our previous point, the critical area in a natural theology is not the interplay between the human and divine, but the work of the Holy Spirit.

Responding to Barth's impetus, it is clear that the doctrine of natural theology needs to have a theological core, and for us, that is the Holy Spirit. For most systematic theologians of the last century the Holy Spirit has not been as important as the first two members of the Trinity, the Father and the Son. It is not surprising, therefore, that the Holy Spirit has not played an important role in many elucidated natural theologies. But this is a weakness, and one that cannot stand if natural theology is to move forward into the 21st century. In his exploration of a pneumatological theology of religions, Amos Yong notes the importance of pneumatology: "It is precisely because the Spirit is both universal and particular, both the Spirit of God and the Spirit of Jesus the Christ, that pneumatology provides the kind of relational framework wherein the radical alterity – otherness – of the religions can be taken seriously even within the task of Christian theology." [72]

The contemporary theologian who, perhaps, has done the most work on the Spirit is Jürgen Moltmann. In his work, *God in Creation*, Moltmann offers keen insights into the work of the Spirit in creation, and can be a dialogue partner for us. Moltmann notes, "without a pneumatological doctrine of creation there cannot be a Christian doctrine of creation at all. But cosmological reasons are cogent also: for without a perception of the creator Spirit in the world there cannot be a peaceful community of creation in which human beings and nature share." [73] Moltmann argues that the role of the Spirit in creation has been pushed out of the theological debate by a mechanistic world picture. In contrast, he claims:

1. The Spirit is the principle of creativity on all levels of matter and life.
2. The Spirit is the holistic principle.
3. The Spirit is the principle of individuation.
4. All creations in the Spirit are in intention "open". [74]

Only as the role of the Creator Spirit is re–established can we begin to accentuate the immanence of God and consequently, a natural theology that has its roots not in humanity, but in the divine activity. General revelation will then provide a God–given (through the Holy Spirit) point of contact.

5. A distinction needs to be made between natural theology and a theology of nature.

By rejecting a completely autonomous natural theology (see point #2), we in no way want to devalue the role of nature or creation. In our first point we spoke to the role of the divine in offering his revelation to the world through creation. Here we want to highlight the key role that creation plays in revealing God, hence, "a theology of nature". Calvin, as we have noted, recognized this, as does Colin Gunton:

> To begin with a general point, may we not say that the world reveals the glory of God because it is creation: that its contingent patterns of being have their own intrinsic rationality; that is to say, they reveal the hand of the creator because of what they are in themselves, not because they are routes to proving the existence of God? Thus the world reveals its maker not because it is continuous with God, but because it is distinct, different even... Then there is a second point, though it is related to the first. The world reveals the being of God by virtue of its capacity to be a framework for culture, that is to say, for all the varieties of human thought, action, and art that take place within it. The fact that there can be culture is a mark of the world's createdness: "God saw all that he had made, and behold, it was very good." In a sense, what I am suggesting is that secular features of the world's being are as relevant to our concerns as those that are apparently religious: good art as much as "religious" art. Those things that are discontinuous from the divine speak the power and deity of God by virtue of what they are. [75]

This is simply a way of noting the valuable effects of general revelation within nature or the world. These "imprints" of the creator need to be recognized for their crucial role in displaying God's revelation.

6. An emphasis on natural theology will open up new vistas in theology.

This relates directly to our work in this book. If God has in fact been working in other places besides the Christian church, how are we to record and analyze that data? This analysis cannot happen purely in the realm of missiology, or a history of religions approach, but must be enfolded into a cohesive, comprehensive Christian theology. Pannenberg states it well,

> When, therefore, we refer Paul's statement about the knowledge of God from the works of creation to the religions we cannot conclude that they are all from the root up no more than idolatry. In them there is knowledge of the true God from creation, though again and again, of course, there is also the exchanging of the incorruptible God for creaturely things (Rom. 1:23, 25). The one–sided exposition of Rom. 1:19–20 solely in terms of the natural theology of the philosophers has contributed to a one–sidedely negative assessment of non–Christian religions in the history of Christian theology. Today we have to correct this false development and arrive at a more nuanced judgement on the world of the religions. [76]

7. There needs to be a better symbiosis of "natural" theology and "revealed" theology.

While not advocating a complete merging of these two categories into one, nevertheless we need to consider ways that these two (at times) competing worldviews can mesh together. The continuing separation of the two creates a bifurcation and subordination that is unintended in view of the reign of God. Jürgen Moltmann gives us a model of a more symbiotic, holistic doctrine.

Moltmann suggests that *theologia naturalis* is the one theology under the conditions of the *regnum naturae*. Nature is creation–in–the–beginning, so *theologia naturalis* is "theology under the conditions of the pristine creation, and of the human being as the pure image of God." [77] *Theologia revelata* is the one theology in the *regnum gratiae*. "It presupposes the self–revelation of God in the history which is determined by human sin and death. Consequently the full concentration and expression of this theology is the *theologia crucis*." [78] *Theologia gloriae* is the one theology in the *regnum gloriae*. This is the "universal fellowship with God which knows no limits". [79] In these three areas Moltmann sees one single theology under different times and conditions; the preceding theology is absorbed by the one that comes after it.

> Theology in the kingdom of glory fulfils the promises and beginnings of natural and historical theology. But it also means that the theology which is possible at any given time is the whole of theology. Revealed theology is natural theology in the conditions of history, just as theology in paradise was revealed theology in the conditions of the pristine creation. The theology of glory is then true natural theology and perfected revealed theology in the condition of a consummated creation and history. [80]

Here, then, is a scheme that enfolds natural and revealed theology in one theology, that of a theology of glory, which represents Moltmann's eschatological thrust. We do not want to suggest, as Moltmann's theology may imply, that the era of natural theology is over, subsumed in a revealed theology. Natural and revealed theology must continue to coexist side–by–side. Nevertheless Moltmann has creatively related and integrated natural and revealed theology into a symbiotic whole.[81]

8. The role of sin, readily alluded to in the context of natural theology, needs to be articulated in the context of general revelation. Just as humans can be blind to special revelation, why can they not be just as blind to general revelation? The neat bifurcation that general revelation (and thus, to a degree, natural theology) is readily available to all, and special revelation only discernable by a few, is too simple. Gunton cites Calvin regarding the limited capacity of the human in general revelation, and consequently, how general revelation needs to be complementary with special revelation:

> As Calvin argued, there can be a general revelation that is ignored through human blindness and capacity for the creation of idols. The creation may reveal the glory of God, but "not one in a hundred" recognizes it for what it truly is. There has, then, to be a distinction between a general revelation in nature, which is indeed there, and the human capacity to appropriate it... That is the point of Calvin's view that without the Bible as a pair of spectacles, we are unlikely to be able to recognize general revelation for what it is. The doctrine of general revelation is not therefore something that operates in parallel with biblical revelation, but is derived from it. [82]

If this type of natural theology – complementary and subordinate to special revelation, yet keenly significant as general revelation – is elucidated, then the necessary outcome will be a flowering of source material not only from the psychological and sociological worlds, but also from the religious contexts of the world. Natural theology becomes a prime resource, theologically speaking, that enables us to move forward and examine material in other religions.

In conclusion, by bringing to together these epistemological (postmodernity and postfoundationalism) and theological considerations (natural theology), we can begin to see how we can proceed towards other religions. By adopting a holist epistemology, and by putting forth a rightly conceived natural theology, we can begin to explore what other religions are saying, and incorporate them into a framework of Christian pluralistic theology.

CHAPTER THREE

A Christian Doctrine of Creation: Two Representative Models

In our first chapter we noted the pluralistic context of our time, and the impact that should have on our theology. In chapter two, we then developed an epistemological rationale out of postmodernity in holism. We also crafted a theological rationale in natural theology, a clear appropriation of which can open us to more of God's working in other religions. Here our intent is to look at two representative contemporary theologians who have focused on a doctrine of creation in their writing. One is a post–Barthian – Colin Gunton – the other is a pluralist – Ninian Smart. Our intent is not to show a full–orbed doctrine of creation at this point, but rather to set the stage for further inquiry by bringing some of the seminal areas in the doctrine of creation to the fore, as well as studying the implications that arise from doing or not doing theology from a pluralist perspective.

Colin Gunton

Colin Gunton has written a number of perceptive works, focusing notably on the doctrine of creation in two works, *The Triune Creator* (1998) and *The Doctrine of Creation* (1997). A number of aspects to Gunton's work concern us here, namely his Trinitarian focus, the context in which the doctrine of *creatio ex nihilo* emerges, and the christological mediation of creation. Each of these three aspects has particular relevance in examining a doctrine of creation in a pluralistic environment. Gunton's will be an example for us of a post–Barthian, constructively Christian doctrine, which is not in dialogue with extra–Christian sources.

Gunton is rigorously Trinitarian, as evidenced by the title of his historical and systematic look at the doctrine of creation, *The Triune Creator*. In his opening chapter of that work, Gunton argues that a "trinitarian theology of creation makes it possible to understand that the creation remains in close relation to God, and yet is free to be itself .... Together the christological and pneumatological structuring of the doctrine provide a ground for the knowledge of both creator and creation as they are both in themselves and in relation to one another." [1] This Trinitarian focus means that at every turn Gunton attempts to relate the systematic aspects of a doctrine of creation to the members of the triune godhead, though we would note that Gunton's triune focus leans in particular on the Father and the Son. In fact, Gunton has argued that the doctrine of the Trinity, with its epistemology of the one and three, provides a needed answer to the transition from modernity to postmodernity, particularly in how the universal and particular relate together. A Trinitarian doctrine of creation helps us hold together both the particularity and universality of humanity's needs.[2] For Gunton, the three members of the Trinity are to be maintained as the grounding principle through which discussion about relatedness, the universal and particular, and meaning – issues central to pluralism – can take place. [3]

Gunton carries this stance into an essay on the doctrine of creation and pluralism, where he posits that "the plurality in unity of the triune revelation enables us to do justice to the diversity, richness, and openness of the world without denying its unity in relativist versions of pluralism." [4] A Christian doctrine of creation, for Gunton, allows for an affirmation of reality and meaningfulness for the *entire* world, against the fragmentation somewhat inherent in postmodernity. As such the doctrine of creation is key in providing "a common foundation for all the human enterprises we call culture – not just theology, but science, politics, ethics, and art as well." [5] For Gunton, creation has been vital to establishing the *unity* of creation.

Consequently, Gunton does not have much use for a natural theology; instead he argues for a theology of nature, in which God is the source of all things: "A theology of nature is the gift of biblical revelation, for it teaches us that the unity of things is upheld neither by the formal causality of the Greeks nor by the supposed omnipotence of human reason, but by the incarnate Lord whose power was exercised in the power of the Spirit and in weakness." [6] Gunton implies an adherence to general revelation without supporting natural theology. God is at work in the world, with the Trinity providing the crucial role in establishing both the distinction between creator and creature, as well as the goodness of creation. However, Gunton's statement, "the trinitarian formulation of a doctrine of creation … allows God to be God, the world to be the world, distinct beings yet related by personal mediation as creator and creation," [7] though strong on delineating between creator and creature, but does not show adequately enough not only how creation can be related via mediation but also how the distinctives of creation can be maintained. For example, in a section on providence, Gunton suggests that we conceive of providence chiefly in terms of two models: "the Son as the giver of structure, and the Holy Spirit as the one who gives the world space to become within but not apart from that structuring. The hinge on which all turns, so often neglected in theologies of creation, is the incarnation." [8] There is no explanation of how the world "becomes" within that structuring or how the particularity is engaged, realized, or nurtured. Gunton argues in another work that the Spirit is the key for particularity.[9] However, his exposition makes clear his concerns are issues of personal existence and life within the church. There is no mention of the wild diversity inherent in comparing a New Guinean polytheist with a Tibetan monk, or an Islamic fundamentalist with a Southern Baptist.

Where Gunton falls short is in explicating the diversity of creation via the Trinity. He constructs a strong case for particularity in creation, and divine unity in creation, but there is no corresponding divine plurality in creation. For instance, where and how is the divine plurality seen elsewhere in creation? What is the Spirit's role of particularity around the world? Here Gunton's insistence on a theology of nature does not serve him as well as the kind of natural theology we have suggested might – one rooted in the Spirit and complementary to special revelation. For example, the diversity of creation can be accounted for by virtue of the Trinity through the ongoing work of the Spirit rather than a theology of nature. If the world is appreciated primarily on its own merits, then it is subject to the same Barthian critiques of Brunner, wherein the *imago dei* cannot be seen as the point of contact between humanity and God because of the fallenness of human nature. What is needed is a divine basis for the plurality and diversity of nature – a spirit–driven natural theology. Thus the key aspect once again becomes theological: the divine plurality in creation is so because of theological reasons, not natural ones.

The second aspect of Gunton's program that concerns us is how the doctrine of creation emerged in its Greek milieu. Against the cosmologies of the Greeks, starting with Heraclitus and Parmenides, and continuing through to Plato and Aristotle, a biblical doctrine of creation engages pantheism. Gunton puts it thus: "Is the universe in some way divine, in the sense that it accounts of itself for the way that it is, or is it the creation of an agent who is other than it, and, specifically, personal?" [10] Pantheism has continued in a modern form through the rise of science as a replacement and total explanation of the mystery of the universe, Gunton argues. This bifurcation of theology and science is not because the two are inimical to each other, but rather is due to the inescapable tension between a Biblical and Hellenic world view: "Either the world creates itself, or it is the product of a personal creator." [11] By positing a personal creator, a Christian doctrine of creation has differentiated itself from any pantheistic conceptions, and is thus to be retained in any pluralistic constructs.

This aspect of a doctrine of creation also applies to pluralism. The doctrine of a personal creator (and its derivative, *creatio ex nihilo*), as developed in a biblical cosmology, both affirms its own merits, as well as distinguishes itself from other possibilities. As we shall see as we move forward in our project, the doctrine of a personal creator is central for Christianity in the face of competing cosmogonies from other religions. The oft–repeated catechism, "I believe in God the Father, maker of Heaven and Earth," points to the role in the Christian tradition of a personal creator, as opposed to, for example, the impersonal demiurge of Plato's *Timaeus*. [12] Likewise the doctrine of *creatio ex nihilo*, not only defines the actions of the creator, but also puts limits on creation: "In particular, biblical characterizations of God's freedom to act in and over against the world require an absolute qualitative distinction between creator and creation. That is the point of the teaching that creation is 'out of nothing': a creative act in the purest sense of the word, in which God brought it about that, when there was 'once' nothing but God, there is now God and a world other that he." [13] As we begin to examine other religious belief systems we shall see how important this doctrine is in addressing forms of pantheism, panentheism, and deism.

The final aspect to Gunton's treatment of the doctrine of creation is really his main emphasis, namely, the christological mediation of creation. Showing how a theology of mediation arose, chiefly through Irenaeus and Origen, Gunton contends that just as a Biblical cosmology argued against pantheism, so too a doctrine of personal mediation through Christ argued against any other type of mediation (i.e. Plato's forms). Using Irenaeus' imagery, Gunton shows how God creates by means of his two "hands", the Son and the Spirit, which means he is "unlike the deities of the Gnostics and the One of neoplatonism in that he does not require beings intermediate between himself and the world in order to achieve his ends.... This is accordingly a theology of mediation which breaks through Hellenic doctrines of degrees of being." [14] These two hands are crucial for Gunton to 1) maintain the Trinitarian aspect of creation; 2) stress the christological import of the doctrine, drawn chiefly through New Testament texts; and 3) highlight the differentiation between creator and creature that a theology of personal mediation acknowledges in contrast to a graded chain of being.

Having already discussed the first feature of Gunton's two hands' emphasis, the second and third also require attention. There can be no mistaking the christological element of creation that the New Testament writers accentuate. John 1:1, alluding to Genesis 1:1, is well–known, as is Colossians 1:16, where Jesus Christ is the one in whom "all things are created, in heaven and on earth ... all things were created through him and for him. He is before all things, and in him all things hold together." The emphasis is so striking that it does seem that theologians who attempt to discuss a doctrine of creation solely on the basis of Genesis 1 and 2 may miss the full Trinitarian thrust of the doctrine. Regarding the third feature of Gunton's theory of mediation, the platonising effect of placing intermediaries between creator and creation denies the emphasis of two distinct realities – creator and creature. By positing an intermediary that is not God, a third class of beings is introduced, as is a lower–to–high sense of reality. This platonising leaves the traditional triune God behind and rushes forward to pantheism, or at least, panentheism. Again, here Gunton is attempting to recover a truly triune sense of creation, an idea that alone can adequately represent the biblical sense of the doctrine:

> Tradition "attributes creation to the Father, salvation to the Son and life in the church (etc.) to the Spirit. That encourages modalism and draws attention away from the New Testament affirmations of the place of Jesus Christ in the mediation of creation as well as of salvation.... It should be said that creation, reconciliation and redemption are all to be attributed to the Father, all realized through the work of his two hands, the Son and the Spirit, who are themselves substantially God. There is mediation, but it is through God, not ontological intermediates. This tends to be lost when any other mediator conceived independently of the 'two hands' – Platonic

forms, Aristotelian causes, Lockean or Newtonian substance, Berkeleyan archetypes – becomes the central focus of attention." [15]

For a doctrine of creation to be adequately discussed, the role of Christ as mediator of creation and salvation must be included. Leaving Christ out of a dialogue on creation is simply not possible for Gunton.

This approach, however, has its weaknesses. The first may be that Gunton has allied himself to the objects of Feuerbach's critique – his concern with creation is mainly Christ, and not creation. Secondly, his strong christological, and perhaps christomonist, doctrine of creation does not allow the doctrine to stay in the realm of creation without moving to soteriology. In another article on creation, Gunton's comments on the role of the Spirit are preceeded by this introduction: "We may explore the theology of the Spirit only through the lens provided by Jesus Christ." [16] This single–minded focus may preclude him from being open to findings and insights from general revelation.

In the end, Gunton is representative of the majority of Christian theologians who are not in dialogue with sources outside of the Christian tradition. While opening up the possibility of a 'theology of nature', Gunton nevertheless is reluctant to develop any kind of natural theology, and as a result has limited his theological sources, I believe. To be fair, Gunton conceives of his work in *The Triune Creator* as more of a historical study, and as such, there is little to discuss in the history of Christian doctrine on this systematic issue that dialogues with other religious sources (we will examine this more closely in chapter four). However, I suggest that the definition of the other word in his subtitle – systematic – be expanded to include not only looking at a doctrine in terms of the Old and New Testaments and theology as a whole, but also to source material in general revelation. This would make his systematic theology truly "systematic" in that it would examine all of God's working in the wider world. This Gunton has not done.

Further, his underemphasis on the "third hand" of the Spirit has left his creation theology not as fully–orbed as it could be. We have already noted his focus on Christology, to the exclusion of Pneumatology. Gunton argues that the Spirit's actions must not be separated from that of the Father and the Son, in order to "avoid the worst kind of anthropomorphism – of making the Spirit a kind of individual agent in relation to the created order." [17] While this is true, he remarks that the dogmatic tradition does not give as much assistance in this area! Rather than quoting a few Cappadocian sayings, perhaps

Gunton needs to develop that area himself – be *more* Trinitarian. A fully developed pneumatological study of creation will be more attentive to revelation outside of the Christian tradition. At any rate, what Gunton has done is to provide a solid reference point in constructive Christian theology, outlined the major Christian systematic positions in creation, and shown how the Christian doctrine was created as a statement against other (competing) worldviews.

Ninian Smart

Our second theologian on creation to consider is Ninian Smart. He addresses this issue from a decidedly non–traditional viewpoint, one that would be considered pluralistic, described in his own words as *neotranscendentalism*. Neotranscendentalism is a "transcendent perspective [that] has to be presented in the new context of scientific thinking and humanistic enquiry" and is explained as thus:

> It is intended as Christian, because we are loyal to the Christian past (or pasts); it is sensitive to the emergency of a world civilization, so it is not parochially Europocentric; it includes an interpretation of world history, not just the 'salvation history' from ancient Israel to the modern Church; it is set within the bounds of modern epistemology, and combines soft non–relativism with possible inner certitude; and it reflects its transcendent focus, and so is critical of flat materialisms and humanisms. [18]

Smart later sums up his position as "neotranscendentalist, pluralistic, social Trinitarian, universalist panentheism embedded in a soft epistemology." [19] There are two reasons for Smart's inclusion: his avowedly pluralistic stance, and his inclusion of both a doctrine of creation and Trinitarian analysis. Regarding the former, though he advocates a pluralistic perspective, Smart nevertheless defines his position as "soft–non–relativism", which is seen as soft "because of the absence of proofs and of decisive evidence, and because of the flexibility of organic religious systems; and non–relativist because we do not draw the relativist conclusions from that fact." [20]

In analyzing Smart's doctrine of creation, which he calls "divine creativity", we must recognize how the creativity takes place. In describing how creation emerges, Smart writes, "As the blessedness of agape consists in self–giving, the Heavenly Father desires to create other self–conscious beings to share in the joy of the Divine Life by being ontologically united with his Son, within the love–bond of the Spirit. This she does through an emergent evolutionary and historical process which the Divine Son enters during his human embodiment."[21] This "emergent evolutionary and historical process" includes various levels of "consciousness", from material energy to the human race, all of which have been brought into being by the continuous creativity of God, the Trinity, which alone is ontologically independent.

This emergence takes place primarily via the participation of the Divine Life in human history, seen through the notion of the Cosmos as God's Body. Smart amplifies this through several points:

1. We are as part of God's body already sacramentally sharing in her life.
2. Since the Divine is working in all things then she is everywhere: we are forever in the presence of the Divine, so that whichever way we turn we find God.
3. This in turn means that churches, pilgrimages, sacred places, etc., are in essence no more where God is than is anywhere else....
4. We can see in other faiths some of the same spiritual message – in Hinduism because, obviously, of Ramanuja's use of the analogy; in Taoism, for the Tao is the divine principle animating the cosmos; in many small–scale religions with an acute sense of the presence of Spirit in the things and forces around and within us. [22]

Here is a strong emphasis on immanence, and perhaps even a full–scale panentheism, one that has echoes of Jürgen Moltmann's chapter in his *God in Creation*, entitled "Embodiment is the end of all God's works". The difference between the two is seen in that Moltmann's concern is to reverse the tendency to make the soul paramount over the body. [23] Smart's interest is the opposite, as we see almost a fully developed Platonism whereby the body or incarnation is only necessary as a teaching tool:

> The material processes of the cosmos are, moreover, the veil which protects the creatures from the full glory and even wild force of the Divine Being. This veil is also the means whereby conscious beings gain their freedom. It has salvific significance, in that without incarnation in bodies the living creatures could not learn their powers, and so gain their individual exercises of freedom. [24]

In fact for Smart, though he retains the embodiment of Christ as necessary (*contra* Hick's argument that divine embodiment has unacceptable implications for pluralism [25]), his Christology displays so many modalistic tendencies, it is difficult to accept his "social trinitarianism" at face value.

Regarding this immanence, Smart holds a key place for embodiment, not only in a spacially constitutive way, but also through the concept of evil, where God emotively participates with her creation: "The creation of the galaxies and sentient beings causes the Trinity sorrow: for how could she create a cosmos with free beings unless she was willing to come down and be an *avatara* and so suffer among us?" [26] This unity *with* creation results in a unity *of* creation for Smart, as humans and other beings are not so conceptually divided. [27] This unity with creation is seen in a paramount way in a "threefold kenosis":

> First, as *antaryāmin* of the cosmos as a whole God guides the forces of the world but by the very fact that she contracts to be regular, to provide a stable background for the free ranging conscious animals that she knows that the cosmos will grow, there are limitations on the ways she is willing to act (like removing the power of fire because someone is just going to be burnt). Second, she leaves freedom to conscious beings to learn from nature and struggle through to their own identities, and so does not dominate the individual from her position behind consciousness. Third, as Christ she is self–emptying in the way Jesus has abandoned omniscience and omnipresence and other divine attributes but really enters into the life of the conscious individual in that cosmos which veils the Father from him (and yet he becomes intimately aware of him). [28]

In many ways, Smart anticipates the kind of "other–religion" informed Christian theology that we are seeking, and will in fact elaborate later.

Nevertheless, we are left with Smart's leaning toward panentheism, which is the often–repeated by–product of a consciously pluralistic theology. In attempting to take into account other religious viewpoints, many theologies succumb to the lure of panentheism, which we can define as the view that God is in all things. Smart is no exception; but, later in this book, we will laud his efforts as being one of the few that uses material from other religions in developing a Christian theology. [29] Here we must focus, though, on the upshot of Smart's approach: Whereas Gunton's particularist approach has produced a fairly insular Christian doctrine, Smart's pluralist approach has produced a doctrine that falls prey to the achilles heel of pluralism, namely that it tends to reduce the incredible diversity of different religions to one watered–down religion. What this approach achieves in terms of vitality of sources is offset by its adoption of the hegemony of modernity in its approach. As we have already noted, pluralism, in this form, retains little of the distinctiveness of the religions while offering a single solution as a replacement for the diversity it was attempting to serve. This tendency towards Platonism and against the particular illustrates the difficulty with any pluralistic "system". This is why Gunton's desire – to bring the one and many together – is so crucial for a postmodern and pluralistic theology. Whether or not Gunton's scheme achieves that is another matter. Smart's advocating of a spiritual reality, or *darsana,* which is available to all, is also an initiative whose supposed success is tenuous. Yet at least he, without developing any methodology in natural theology or general revelation, has attempted to go outside of the Christian box. And for that, I believe, his theology is the richer. He is making use of general revelation, and consequently, his doctrine is open to a wider audience. [30] It remains to be seen whether there can be a third way – a doctrine that develops from an explicitly Christian position, takes into account material from other religions, and does not slip into the vague, neutered modernity of pluralism. This will be what we will develop in our final chapter.

But before we explore that third way, it will prove useful to take another turn on our journey to see if we can find any resources from the historical Christian tradition to guide us. We have developed a theological rationale – natural theology – but has that rationale been used effectively throughout the history of theology? Has the Christian tradition used material from other religions in developing its theology, and if so, how? We will find some surprising answers that will lead us, in turn, to explore a biblical rationale as well. With these two explorations completed, we will then examine cosmogonies and cosmologies in Buddhism, Hinduism, and Islam to search for material to be incorporated into a Christian doctrine of creation. In our final chapter, we will then integrate these findings in three areas in beginning to construct a contemporary pluralistic Christian theology of creation.

CHAPTER FOUR

How the Christian Tradition Has Used Material from Other Religions

The Christian church has a long history. Buddhism, Hinduism, and Islam have long histories as well. It was inevitable then, that these histories would cross from time to time. At these intersections of people, cultures, and ideas, theology has arisen – theology that is sometimes cognizant of differing religious worldviews, sometimes not. Christian theologians have attempted to discern if and how these differing religions fit together, or, what the implications might be theologically, given the alternative of diffuse ways and ends. Overall, the church has struggled with how to relate to its religious neighbors, something that will be reflected in this section. Many have tended toward isolation and insulation from other religions, others have embraced pluralism, and others have fallen somewhere in the middle.

How has the Christian tradition used material from other religions? That is the question before us. Given the insights that we find in other religions, has the Christian tradition used these as source material in its own theologies? In asking these questions, we are also tangentially asking two other questions – what has been the Christian tradition's attitude toward other religions, and, in the case of a positive response, how has it gone about interacting with that material? To answer these questions, we will take a survey of Christian theologians' attitudes toward other religions. Our method will be to take a sampling of thought from major theologians from representative eras in church history. We will consider more examples as we move into the 20th century simply because more theologians have been working on the task. My desire is not to present a history of the theology of religions, but rather an analysis of the usage of doctrines from other religions. For if in fact we have established that a theological rationale is present in natural theology, we must look to see if any theologians have acted on that impulse and used material from outside of the Christian milieu for their theologies. This section will provide enough analysis for us to observe what has happened (or not happened!) in the Christian tradition, and give us resources to move forward in our endeavor of gleaning all that we can from God's worldwide revelation. By way of introduction, we may state that in general there is a pattern among our sampling of acknowledging revelation being present in other religions, but of not following through and using or analyzing that material. Our contention is that the reasons for this have to do with the unique cultural situations each theologian found himself responding to.

Early Church & Augustine

We will begin with the early church, and from the outset we must note that the attitude of the early church towards other religions was not one of accommodation. Coming as it did out of Judaism, the early church was more concerned to create its own identity, not only defining itself compared to Judaism, but also defining itself over against other religions. Consequently, it did not view other religions and philosophies as capable of being harmonized with the Christian story. Consider Justin's claim that pagan gods were the offspring of fallen angels and human women and the dire consequences thereof:

> But the angels transgressed, and were captivated by the love of women, and gave birth to children who are called demons. Afterwards they subdued the human face to themselves, partly by magical writings, partly by the fears and the punishments they occasioned, and partly by teaching them to offer sacrifices ... which they needed after they were enslaved by lustful passions. Among men they sowed murders, wars, adulteries, imtemperate deeds and all wickedness. The poets and mythologists did not realize that it was the angels and the demons begotten by them who did these things ... and so they ascribed them to God himself, and to those reckoned to be his offspring. [1]

A somewhat more positive theory of pagan religion held that it was part of God's plan to lead humanity towards its goal in Christ. Clement of Alexandria writes, "Accordingly, before the advent of the Lord, philosophy was necessary to the Greeks for righteousness.... For this was a schoolmaster to bring the Hellenic mind, as the law, the Hebrews, to Christ. Philosophy, therefore, was a preparation, paving the way for him who is perfected in

Christ." [2] But by and large, the church saw other religions as often demonic, and certainly not to be mingled with. Gerald Bray, in a fascinating article entitled "Explaining Christianity to Pagans" sums up the situation,

> In the final analysis, therefore, the Apologists were not forerunners of the kind of religious dialogue advocated by pluralists in our own time. However much they may have agreed that all human beings possessed something of the divine Logos, and even when they were prepared to commend a man like Plato for having made the best use of the Logos given to him, pagans were still the victims of demonic deception, with only a partial (and usually misleading) grasp of the truth .... As far as Justin was concerned, it was baptism that underscored the fundamental character of the Christian, as one who was born again in Christ. Such a person had the spiritual equipment necessary to distinguish the true from the false in paganism, but in making this distinction he was only bringing out the blindness of the Greek philosophers even more clearly. For when all was said and done, the philosophers lacked the spiritual perception needed to distinguish right from wrong, and so were corrupted even in their better moments. This was a far cry from accepting that paganism had something positive to offer, or that it was possible to become a Christian without renouncing one's former life. Christ came into the world to judge paganism, even in its noblest forms. Whatever differences of approach there were among the Apologists, their basic message was the same – only by repentance and rebirth could a human see the kingdom of God. [3]

It was from this milieu that one of the great theologians of the church arose, Augustine (354–430), whom we will take as a representative of the early church. It is difficult to summarize such an important and prolific figure as Augustine. His thinking on the relationship of the church to the state, the Trinity, original sin, and his devotional writing all profoundly influenced later western Christian thinking, especially the magisterial reformers. Latourette writes that "no other Christian after Paul was to have so wide, deep, and prolonged an influence upon the Christianity of Western Europe." [4] Specifically, no one has had a greater influence in the area of dogma. It is particularly in his struggles with various groups – Manicheans, Semi–pelagians, Donatists – that Augustine developed his thinking. Portalie's description of Augustine's work is telling: "His twenty–year struggle with Donatism resulted in a complete dogmatic system on the Church, the great work and mystical body of Christ and the true kingdom of God, on its part in salvation, and on the intimate efficacy of its sacraments. The Manichean controversy forced him to explain clearly the great problems of the divine Being and the nature of evil."[5] Augustine struggled to define Christianity and the church *over against* those whom he saw as heretical. Heresy was something to be rooted out, to be fought against, and in this we see Augustine's link with the early church fathers, who fought to carve out and preserve this new religion in a diverse world. We can, in fact, understand most of Augustine's work as being composed as "polemics against heretics." The idea of incorporating competing worldviews within Christianity would not have occurred to Augustine. His theological intention was directed elsewhere.

However, having noted this, we must also comment on three factors in Augustine's work that force us to acknowledge that Augustine is not as closed off to non–Christian sources as we might at first think. The first is Augustine's clear use of Greek Platonism. Augustine's indebtedness to Platonic theories in the areas of philosophy, theodicy, nature of the created world, etc. is clear, leading some to wonder if Augustine really adapted Christianity to Neoplatonism. [6] However, Augustine's own words on the subject provide a fascinating insight into his method. In *Against the Academics* he writes that there are two methods of arriving at the truth, authority and reason: "Now in the matter of authority I have chosen Christ for my leader, from whose direction I will never deviate." As for philosophy, he adds, "As regards matters which are to be investigated by close reasoning, I am of such a mind that I impatiently desire to grasp the truth not only through faith but also by understanding. I am confident that I will find in the Platonists nothing that will be at variance with our sacred literature." [7] Here Augustine is open to the thought of a non–Biblical philosophical tradition, the adapting of which to the Christian story was not problematic for Augustine. However, the hierarchy is clear – Christ is the leader, and other thought may be incorporated so long as it is not at variance with Christian orthodoxy. The point may be made that adapting pagan philosophies has always been easier for Christian theologians than adapting pagan religions.

The second point to note is Augustine's own sense of using the "spoils" of the Egyptians, which he comments on in *On Christian Doctrine*:

> Moreover, if those who are called philosophers, and especially the Platonists, have said aught that is true and in harmony with our faith, we are not only not to shrink from it, but to claim it for our own use from those who have unlawful possession of it. For, as the Egyptians had not only the idols and heavy burdens which the people of Israel hated and fled from, but also vessels and ornaments of gold and silver, and garments, which the same people when going out of Egypt appropriated to themselves, designing them for a better use, not doing this on their own authority, but by the command of God, the Egyptians themselves, in their ignorance, providing them with things which they themselves were not making a good use of; in the same way all branches of heathen learning have not only false and superstitious fancies and heavy burdens of unnecessary toil, which every one of us, when going out under

> the leadership of Christ from the fellowship of the heathen, ought to abhor and avoid; but they contain also liberal instruction which is better adapted to the use of truth, and some most excellent precepts of morality; and *some truths in regard even to the worship of the One God are found among them* {italics mine}. Now these are, so to speak, their gold and silver, which they did not create themselves, but dug out of the mines of Gods providence which are everywhere scattered abroad, and are perversely and unlawfully prostituting to the worship of devils. These, therefore, the Christian, when he separates himself in spirit from the miserable fellowship of these men, ought to take away from them, and to devote to their proper use in preaching the gospel. Their garments, also,– that is, human institutions such as are adapted to that intercourse with men which is indispensable in this life, – we must take and turn to a Christian use. [8]

This remarkable passage contains some insights that would not be acted upon for many centuries, especially in regards to statements about God. We have already noticed Augustine's usage of Plato, but it appears that he did not follow the full implications of his own suggestions in borrowing from other non–Christian sources. We would also remark here that the tone of using the "spoils" of the Egyptians (Ex. 3:22) does not seem as respectful as our current interest in learning from the "other" demands. However, Augustine was writing in a different time and place, and his paramount concern was for the building up and establishing of the Christian faith. Nevertheless, Augustine declares himself open to using wisdom from other belief systems that he could adapt to Christianity.

The final aspect to notice regarding Augustine's work on heresy is his pastoral bent, something we see in his controversy with the Donatists. The Donatist schism was based on the idea of purity – whether one should remain in a church that has sinners, or pull out to form a still purer community. Augustine argues that the unity of the church of Christ is so fundamental that he is willing to tolerate less than the ideal Christian – "we should take the utmost pains we can to correct the erring ones themselves; not only watching that they should do no injury to the weak, and that they should be delivered from their wicked error, but also praying for them, that God would open their understanding." [9] Augustine then exemplifies his position with a welcoming and pastoral stance toward the Donatists. In a poignant illustration of his desire for the Donatists to come back "home", he describes how a branch can be grafted back into a tree. He writes, "We feel no jealousy towards them, nay, we embrace them; we wish, we advise, we even compel those to come in whom we find in the highways and hedges." [10] Augustine's *personal* stance towards those of different belief was one of warm dialogue, even as he argued strenuously for the true explication of the doctrine at stake.

Augustine thus shows for us his context, one of developing the church and protecting it from heresy. He was not looking to include new insights from other religions. Yet he was open to dialogue with other sources where it was appropriate and under the leadership of Christ, even to the point of embracing Neoplatonism. Would others incorporate this insight and build on it?

Thomas Aquinas

Thomas Aquinas is significant for us, not only in how he used a pagan philosopher, Aristotle, but also for his theological developments in the area of natural theology, which we have discussed earlier. We must begin with Aquinas' general attitude towards nonbelief:

> Accordingly there are two ways in which a man may deviate from the rectitude of the Christian faith. First, because he is unwilling to assent to Christ; and such a man has an evil will, so to say, in respect of the very end. This belongs to the species of unbelief in pagans and Jews. Secondly, because, though he intends to assent to Christ, yet he fails in his choice of those things wherein he assents to Christ, because he chooses, not what Christ really taught, but the suggestions of his own mind. [11]

Unbelief is the umbrella under which all forms of non–Christian belief (i.e. "pagans and Jews") fall. Those with unbelief cannot be compelled to believe, but it may be necessary to compel (force) those with unbelief in order not to hinder the faith of others:

> Among unbelievers there are some who have never received the faith, such as the heathens and the Jews: and these are by no means to be compelled to the faith, in order that they may believe, because to believe depends on the will: nevertheless they should be compelled by the faithful, if it be possible to do so, so that they do not hinder the faith, by their blasphemies, or by their evil persuasions, or even by their open persecutions. It is for this reason that Christ's faithful often wage war with unbelievers, not indeed for the purpose of forcing them to believe … but in order to prevent them from hindering the faith of Christ. [12]

Unbelief is such a serious threat to the Christian community that the church has forbidden "the faithful to communicate with those unbelievers who have forsaken the faith they once received, either by corrupting the faith, as heretics, or by entirely renouncing the faith, as apostates." [13] However believers firm in their faith may communicate with non–believers for the purposes of witnessing to them. Aquinas even deliberates whether the rites of unbelievers should be tolerated: "Though unbelievers sin in their rites, they may be tolerated, either on account of some good that ensues therefrom, or because of some evil avoided. Thus from the fact that the Jews observe their rites, which, of old, foreshadowed the truth of the faith which we hold, there follows this good – that our very *enemies* {italics mine} bear witness to our faith ...." [14] Aquinas' concerns are to protect the church from heresy, and convince others to join the Christian faith. He is part and parcel of a time where the world is made up of either believers or infidels, and the infidels are either to be converted or eliminated, so as not to contaminate believers. It seems clear that Aquinas would not have contemplated learning from other faiths.

However Aquinas did contemplate learning from pagan philosophy, and this returns us to our earlier suggestion that learning from extra–Christian philosophy has always been more accepted than learning from other religions. We have already reflected on Aquinas' nuanced argument about the relationship between faith and reason; he uses similar arguments in discussing how faith can use philosophy. This lengthy citation from his commentary on Boethius' *On the Trinity* helps us understand *how* philosophy might be used theologically:

> The gifts of grace are added to nature in such a manner that they do not remove it but perfect it. So it is with the light of faith that is infused in us gratuitously: it does not destroy the light of natural knowledge with which we are by nature endowed. Now, although the natural light of the human mind does not suffice for the manifestation of the things that are made manifest by faith, yet it is impossible that what is divinely taught to us by faith be contrary to the things with which we are endowed by nature. For one or the other would then have to be false, and since both come to us from God, God would be to us an author of falsehood, which is impossible...
>
> Now just as Sacred Teaching is founded on the light of faith, so philosophy is founded on the natural light of reason. It is therefore impossible that what belongs to philosophy be contrary to what belongs to faith; it rather falls short of it. It contains, however, certain likenesses of what belongs to faith, and certain preambles to it, as nature is a preamble to grace. And if in what the philosophers have said we come upon something that is contrary to faith, this does not belong to philosophy but it rather an abuse of philosophy arising from a defect in reason...
>
> Thus, therefore, in Sacred Teaching we can use philosophy in a threefold way.
>
> *First*, we can use it to demonstrate the preambles of faith, which are necessary in the science of faith as being the things that are proved of God by natural arguments, e.g. that God exists, that God is one, or similar propositions ...

> *Second*, we can use philosophy to make known through certain likenesses what belongs to the faith, as Augustine in his book *On The Trinity* uses many likenesses drawn from the teachings of the philosophers to explain the Trinity.
>
> *Third*, we can use philosophy to oppose what is said against faith, either by showing that these things are false or by showing that they are not necessary. [15]

This highly developed hermeneutic for the use of philosophy in theology shows how Aquinas could be open to Aristotle in multitudinous ways.

It is well known that Aquinas adapted and borrowed frequently from Aristotle, so much so that theologians write of Aquinas "Christianizing" Aristotle. [16] For example, McDermott argues that Aristotle colored the lens through which Aquinas looked at the resurrection:

> Aquinas's achievement was to harmonize the Christian doctrine of creation with Aristotle's conception of the natural world. In the process Aristotle showed Aquinas that the soul, which for Plato defines the human person, is only a part of a human being and exists by nature in a body. Hence Aristotle gave Aquinas philosophical support for the biblical doctrine of the resurrection of the body – and perhaps the confidence he needed to teach a robust view of the body in an age that demeaned it.[17]

Aquinas also frequently uses his doctrine of analogy, which he learned, in large part, from Aristotle. Brian Davies writes, "The notion that terms can be applied to God and to creatures analogically is one to be found in a number of thirteenth–century authors indebted, in this respect, to Aristotle." [18] But Aquinas did not merely adopt Aristotle in a wholesale manner; Vos has observed, "[Aquinas] does not simply recapitulate Aristotle; rather he silently corrects and enriches his position. He employs Aristotelian principles in contexts and ways Aristotle never used them. In doing this Aquinas was doing something quite similar to what his contemporaries did in the areas of liturgical drama, painting, and the like, when they took Bible stories and treated them as if they were contemporaneous." [19]

Aquinas is then a lucid example of a Christian theologian articulating a hermeneutic for, and then using, a non–Christian source – in this case, pagan philosophy.[20] But Aquinas seemed to have little use for non–Christian religious sources, even though he indicates that "natural knowledge" can be of use to the things of faith. His dialogue partners were philosophers, not people from other religions, and so his theology reflects no serious engagement with other religions' teachings.

Luther and Calvin: The Reformation

As we move to the Reformation, we are met by the two towering figures of Martin Luther and John Calvin. As we have discovered with our other two theologians, Augustine and Aquinas, we see that the Reformers were also products of their times. Consequently in neither case – Luther or Calvin – do we find anything approximating what we would now call a theology of religions. Again, their focus was elsewhere. In particular, both were consumed with explicating the Christian faith in a time of great dissatisfaction with the Roman Catholic church, resulting in what we now call Protestantism. We will begin with Martin Luther.

A perusal of Luther's works will find one mention of Buddha, no mention at all of Hinduism, and antipathy directed against Jews and Muslims. Here are a few examples…

> The Jews, the Turks, and the pope will never reach this conviction [that Christ is God and man], for they are offended by it and resent it. If God had had a book written expressly against Jews, Turks, and other persecutors in defense of this article, it would have been this Gospel of John, which is aimed directly at their wisdom. The Turk looks upon us as nothing but archfools. The Jews and the pope, too, regard us as the greatest dunces for believing that this Person, Christ is God and man. They will not concede that Christ is our Savior and the bread of life. They defame and insult us who believe in Christ, the Son of God. The pope treads him underfoot. He robs Him of His Godhead, enjoining the people to perform good works and to do this and that for their salvation. To be sure, the Turk does have a rather high regard for Christ, granting that He is a great prophet; but he does not admit that He is the Son of God. Our salvation must rest on the fact that we look to Christ and that in Him we have the bread of life. None of them believes this. And they are not only silent about this; they also trample it underfoot. [21]

In this passage, we see two of Luther's theological arguments that are crucial for our understanding of him: 1) Luther's central Christology – for him Christ is everything, and salvation rests in him alone; and 2) the Pope is lumped in with nonbelievers – this too is characteristic of Luther's writings as the Pope is one who is no better than a pagan. Luther isn't very concerned about how Christians are to react and interact with "heathen" (though his frequent use of "Turks" indicates how Germany was beginning to come in contact with the Muslim faith); he is more concerned with how the church can stay pure in Christ, and with mobilizing the nascent reformation movement against Rome. There is not even much discussion about punishing "infidels" or compelling them to faith, because of Luther's preoccupation with Catholicism.

In another telling passage, Luther remarks about the role of reason, revelation, and the light of Christ:

> Reason may raise its light aloft and laud it to the skies. And this light, admittedly, may be ingenious in secular and temporal matters; but under no circumstances should it presume to penetrate into heaven. Nor should we confer and consult with reason in questions pertaining to salvation. For in this area the world and reason are stone–blind; they will always remain in darkness and will never shed any light in that sphere in all eternity. Christ is the only Light; He can and will counsel and help.
>
> If Christians regarded Christ as the only Light and staunchly adhered to that belief, then our views, doctrine, and faith would be uniform, and our sermons would convey the same harmonious message everywhere in the world. Our brethren in the Orient would be of one mind with us .... And if I were to travel to Turkey and encounter a Christian who was discussing Christian doctrine and faith on the basis of Scripture, I would say: "That is also my creed." This accounts for our name "Christian"; we all are named after our Lord Christ. Our name reflects our belief that He is our sole Light, Life, Way, Hope, and Salvation. Others may dub themselves Franciscans, Dominicans, Augustinians, Carthusians, Mohammedans, or papists, or anything else. They are no light; they are sheer darkness. If I were to pile on one heap all the others who style themselves lights, they would still be exposed as nothing but will–o–the–wisps or satyrs, who are seen only at night. They lead a person astray into a body of water, where they drown him, or they may destroy him in stone quarries or lime pits. [22]

It is clear that Luther has no place for any natural theology, any hint of truth in any other religious system, or anything else that would detract from his high Christology. Christian revelation through Christ is to be the only source of knowledge for the believer. Primarily an expositor of scripture, Luther finds no place for dialogue regarding religion. He reveals his intense, imperious nature that was so vital for the establishing of Protestantism. Of course, we must mention that, like many Europeans of that time, Luther would not have come into much contact with people of other religions. Theirs was a Holy Roman Empire, after all, one dominated by the life of the church. So Luther's concerns are not for how to learn from other faiths, but how to purify the one faith that is before him.[23]

We have already noted the other great reformer, Calvin, and his contributions in the area of natural theology. Much more of a systematician than Luther, Calvin gives us clearer statements about those in other religions. Calvin holds a central place for Christ in discussing faith in God:

> Therefore, the common saying that God is the object of faith ... requires to be received with some modification. When Christ is called the image of the invisible God (Col. 1:15), the expression is not used without cause, but is designed to remind us that we can have no knowledge of our salvation, until we behold God in Christ.... For though in old time there were many who boasted that they worshipped the Supreme deity, the Maker of heaven and earth, yet as they had no Mediator, it was impossible for them truly to enjoy the mercy of God, so as to feel persuaded that he was their Father. Not holding the head, that is, Christ, their knowledge of God was evanescent; and hence they at length fell away to gross and foul superstitions, betraying their ignorance, just as the Turks in the present day, who, though proclaiming, with full throat, that the Creator of heaven and earth is their God, yet, by their rejection of Christ, substitute an idol in his place. [24]

While acknowledging that other religions claim to know God the Father, because of their rejection of Christ, Calvin claims they cannot in truth know the Creator. In further discussing the role of Christ the redeemer, he writes that the knowledge of God the Creator is of no avail without faith in Christ the redeemer: "By these words [John 4:22 – "you worship you know not what; we know what we worship: for salvation is of the Jews"] [Jesus] both charges every Gentile religion with falsehood, and assigns the reason – viz., that under the Law the Redeemer was promised to the chosen people only, and that, consequently, no worship was ever pleasing to God in which respect was not had to Christ. Hence also Paul affirms, that all the Gentiles were "without God," and deprived of the hope of life." [25] Calvin holds then, no place for divine revelation coming through other religions. [26]

But here again, we must take a step back from our historical survey to note the degree to which all of our theologians have used secular philosophers, and philosophy. Augustine used Platonism, Aquinas used Aristotle, and Calvin used Cicero (to a lesser degree). This represents a usage of material typically seen as outside of the Christian tradition. It may be said that theologians throughout the centuries have been comfortable using non–Christian philosophers, and their protestations against learning about God outside of Christ notwithstanding, their usage of those sources belies to some degree their own methodology. Would they consider Plato, Aristotle, and Cicero to possess some form of general revelation? It is doubtful that that would be acknowledged. To go further, it is clear that none of the theologians discussed would be ready to use any material from other religions in crafting their own theology.

All of the theologians outlined above presented a strong case against a self–sufficient natural theology. All are willing to say that only special revelation (and particularly Christ) can lead one to a saving knowledge of God. All are also unwilling to note general revelation present in other religions (though Calvin comes close), and as such, a theology of religions, as we know it today, is utterly lacking. Aquinas' thoughts on the role of reason provide the strongest theological rationale for learning about other religions, though he himself would probably have demurred from such a course. In all of these areas, we see the church reflecting her times and her needs in those times. As we move toward the twentieth century, we will see more and more reflection on other religions, to the point that at the end of the twentieth century, theologians are beginning to not only have a theology of religions, not only to dialogue, but to even incorporate material from other religions into a Christian theology. Again, the church can be seen as reflecting her times and addressing her current needs. Given our epistemological and cultural era, dialogue with and incorporation of material from other religions is an appropriate direction for theology to take in these times.

Schleiermacher & Troeltsch: The Move Toward Liberal Relativism

We encounter in Friedrich Schleiermacher a radical departure from the conservativism of the reformers. Schleiermacher, the father of 19th century German Liberalism, is notable for his thinking on religion. In *The Christian Faith,* he defines religion famously, as the "feeling of absolute dependence", one that is not merely emotive, but rather a consciousness of being in relation with God. The true nature of religion is shared by all of humanity, so much so that he writes, "Is there not in all religions more or less of the true nature of religion, as I have presented it to you? Must not, therefore, each religion be one of the special forms which mankind, in some region of the earth and at some stage of development, has to accept?" [27] Schleiermacher discusses the religions and rates them on a scale of development moving from idol–worship, through polytheism, to monotheism. Of the three great monotheistic religions, Christianity is the pinnacle – "the original intuition of Christianity is more glorious, more sublime, more worthy of adult humanity, penetrates deeper into the spirit of systematic religion and extends itself further over the whole Universe." [28]

Here we see one of the first serious attempts at ascribing any revelatory value at all to other religions. This would not technically be called the first theology of religions, because throughout the history of Christian theology, there have been a variety of expressions of theologies of religion, though they would not have been labeled as such. Owen Thomas has categorized historical theologies of religion into six major categories: 1) Truth–Falsehood; 2) Relativity; 3) Essence; 4) Development–Fulfillment; 5) Salvation History; 6) and Revelation–Sin. [29] Most theologies of religion before Schleiermacher would be exemplary of the *truth/falsehood* approach – Christianity is true, all other religions are false. Even the apologists may have attributed something revelatory to other religious worldviews, but it was so minute as to not even warrant discussion; thus their theology of religion would have fallen into this *truth/falsehood* category. But in Schleiermacher we see one of the first attempts at breaking out of that mold, and adopting an *essence* attitude toward religions, that is, that there is one essence in all the religions. All religions may be able to be ranked in value and sophistication, according to Schleiermacher, but they still all possess an essence of religion. A modern example of an *essence* approach toward religions would be John Hick. [30]

As we continue to note the development of attitudes toward other religions during the 19th century, and building on the insights of Schleiermacher, we can also turn to the contributions of Ernst Troeltsch. Troeltsch might be called the father of modern pluralism. He is, according to Thomas, a representative of the *relativity* class of attitudes toward other religions – i.e., all religions are relatively true, mainly because they are each expressions of differing cultures. [31] Convictions as to the absolute truthfulness of Christianity are based on the "evidence of a profound inner experience", only though, *for us*: "It is God's countenance as revealed to us; it is the way in which, being what we are, we receive, and react to, the revelation of God. It is binding upon us, and it brings us deliverance. It is final and unconditional for us, because we have nothing else." [32] Troeltsch continues, "But this does not preclude the possibility that other racial groups, living under entirely different cultural conditions, may experience their contact with the Divine Life in quite a different way, and may themselves also possess a religion which has grown up with them, and from which they cannot sever themselves so long as they remain what they are. And they may quite sincerely regard this as absolutely valid for them, and give expression to this absolute validity according to the demands of their own religious feeling." [33] Troeltsch assigns each religion a value in relation to its own people, and more importantly, to its own historical context: "A religion, in the several forms assumed by it, always depends upon the intellectual, social, and national conditions among which it exists." [34]

Though Thomas assigns Troeltsch to his relativity class, Troeltsch goes on to rank Christianity as one of the higher religions, though in the end, only God can judge. [35] Given this tack, Troeltsch would seem to be more accurately representative of a *development/fulfillment* approach, where Christianity is seen as the highest development of religion. James Luther Adams has suggested a helpful amendation of Troeltsch's relativistic position which is worth noting. Troeltsch is what might be termed "iconic" in his view of the religions. Writes Adams, "[Troeltsch's position] is analogous to the view set forth by Nicholas of Cusa in *The Vision of God* (1453). Here in Neoplatonic fashion God is symbolized by an omnivoyant human portrait that faces directly the various people who simultaneously confront it, each of them experiencing the directness of his gaze. Cusa calls the portrait 'the icon of God.' Troeltsch from a somewhat similar perspective asks the question, 'Who will presume to make a really final pronouncement?' And he replies, 'Only God himself, who has determined their differences, can do that.'"[36] Troeltsch also advocates then what we have described as Hick's position, namely, that all religions are manifestations of the Real: "A truth which, in the first instance, is *a truth for us* does not cease, because of this, to be very Truth and Life."[37] In the end, we may say it is possible to be part of a number of categories concurrently regarding a theology of religion, particularly if one is open to seeing revelation in other religions. Troeltsch's position partakes of the essence, relativity, and development–fulfillment approaches to religion.

Having mentioned here the positive attitudes both Schleiermacher and Troeltsch hold toward other religions, we note again the same curious fact mentioned earlier: While ascribing revelatory truth to other religions, neither Schleiermacher nor Troeltsch go on to either explicate any of that revelation or make an attempt to synthesize it with Christianity. The general shape of the various religions is noted, but not the specific content of each religion. This lack of the specific use of any outside religious tradition in formulating their own Christian theology is true of most later theologians as well, even those that see revelation in other religions. In other words, many 20th century theologians see positive divine revelation in other religions, but hardly any go the next step and attempt to meld that revelation with Christianity. Perhaps this is because they feel that these other religions are inaccessible and possess their own inner logic unavailable to an outsider. There is enough information, however, for a Troeltsch to judge some of these other religions as less–developed than Christianity. But again, they do not seem to take the revelation in other systems seriously enough to include it in their own Christian theology. Whatever the reason, they are unable to act on the implications of their own position. But for these 19th and 20th century theological forbearers, their theological innovation was that they were able to acknowledge truth in another system; it fell to another generation of theologians to explore that revelation further.

Barth & Brunner: Neo–Orthodoxy

We may be getting ahead of ourselves a bit by discussing Schleiermacher and Troeltsch, because they anticipate the contemporary discussion so well. We must first pause to reckon with neo–orthodoxy, and the figures of Karl Barth and Emil Brunner.

We must begin with Barth's pronouncement that all religion is unbelief:

> [A theological evaluation of religion and religions] will always understand religion as a vital utterance and activity of this man. It will not ascribe to this life–utterance and activity of his unique 'nature', the so–called 'nature of religion', which it can then use as a gauge to weigh and balance one human thing against another, distinguishing the "higher' religion from the 'lower', the 'living' from the decomposed', the 'ponderable' from the 'imponderable'....Revelation singles out the church as the locus of true religion. But this does not mean that the Christian religion as such is the fulfilled nature of human religion. It does not mean that the Christian religion is the true religion, fundamentally superior to all other religions.
>
> We begin by stating that religion is unbelief. It is a concern, indeed, we must say that it is the one great concern, of godless man…
>
> Revelation is God's self–offering and self–manifestation.... From the standpoint of revelation religion is clearly seen to be a human attempt to anticipate what God in His revelation wills to do and does do. It is the attempted replacement of the divine work by a human manufacture.... For what is the purpose of the universal attempt of religions to anticipate God, to foist a human product into the place of His Word, to make our own images of the One who is known only where He gives Himself to be known ....
>
> In our discussion of 'religion as unbelief' we did not consider the distinction between Christian and non–Christian religion. Our intention was that whatever we said about the other religions affected the Christian similarly. [38]

What is, perhaps, new in Barth's "orthodoxy" is that he goes on to include Christianity in his critique of all religion. Religion is incapable of bringing us to God. So we have the need, in Barth's system, for revelation, for God to come to us, which He has done supremely in the person of Jesus Christ.

There would not seem to be much room for a positive evaluation of other religions for Barth, particularly in his comments on religion and revelation. [39] In his doctrine of creation, we see Barth further elucidating his position. Barth demarcates his doctrine of creation tightly around the Trinity, and in so doing, leaves little room for any other possible theological (and religious) explanations of the world. Distinguishing between "history", "saga", and "myth", Barth maintains that the creation story is pure "saga", and as such is "non–historical" or "pre–historical". [40] Because of saga's link with history, it is to be distinguished from myth, and this Barth does in a long subsection comparing the biblical creation saga with the Babylonian epic *Enuma elish* and the cosmogony of Berosus (3rd c. B.C.). [41] These other ancient documents are considered mythical texts, while the biblical story is considered "historicized myth", i.e. saga. Saga's linkage in time and space to an external referent, in this case, God, render it a more accurate and 'real' option for belief. We suggest, then, in this context Barth finds that Genesis 1 and 2 is itself a critique on pluralism; Genesis 1 and 2 form a subtle but real backdrop for Barth's opposition to natural theology and other religious options.

Barth tacks in the same direction in his analysis of the creation of humanity later in the first volume of his doctrine of creation. By contrasting the biblical account with both the "Babylonian and other pagan traditions" and the "Greek conception of man," Barth is able to assert the "true and genuine dignity of man." [42] So while Barth's aversion to finding in religion, with its basis in human reason and autonomy, any core revelation is well known, what we find here in these texts is perhaps the beginnings of Barth's thinking on pluralism. Near the end of his doctrine of creation, Barth sums up his perspective, being loathe to give any credit to other world views: "None of the world–views so far known to us has attained to the concept of creation by following to the end the way from noetics to ontology and geneseology, but has usually remained stuck either in noetics or at most ontology." [43] The question of origins, for Barth, establishes the finality and exclusivity of the Christian God.

We have noted in chapter two Barth's debate with Brunner regarding natural theology. Barth's "nein" is still heard in theological quarters today.

Yet, returning to Brunner again for his full–fledged thoughts on other religions, particularly in his *Revelation and Reason* (1941), we see not only a theology of religions, but also a cursory analysis of the doctrines of other religions. This represents a step forward in the development of analysis of other religions. For Brunner, traces of the idea of revelation do occur in other religions, though neither primitive, polytheistic, or mystical religions make claims to a universally valid revelation. [44] As such they are not comparable to the claim of the Christian revelation. However, Zoroastrianism, Islam and Judaism do claim to be based on authentic revelation. While they are impressive forms of religion, ultimately they are forms of rationalistic theism with moralistic overtones, yet lacking redemptive qualities. Of Zoroastrianism: "Thus the form and the experience of prophetic revelation clothes a content which man, in virtue of his moral consciousness, fundamentally has in himself; it is the self–affirmation and the self–confirmation of the natural moral consciousness. In the strict sense of the word, here there can be no question of revelation…. It lacks precisely that which transcends the moral reason, the mystery of generous love, the paradoxical blend of holiness and mercy, which is the core of the message of the Old Testament." [45] Of Islam: "When we look at the core of the doctrine of Islam we see that, exactly like Zoroastrianism, it is a rationalistic, moralistic Theism. It teaches what, according to Luther, reason teaches everyone: that there is a God, that this God demands good and that He punishes evil. Here there is no revelation of the gracious mystery: that the holy God loves the sinner. In spite of all the other elements – gathered from the most varied sources – Islam is not essentially different from the 'religion of the Enlightenment.'" [46] Of Judaism: Regarding the messiah, "they confess, therefore, the temporary character of the revelation that has come to them; but they refuse to admit that the final revelation has taken place…This religion has both significance and dignity … but it has no redeeming power. It is the religion of self–redemption… The point of our present discussion is the fact that here there is no *claim to revelation*, but the claim to the validity of a rational idea." [47] Not unlike Schleiermacher and Troeltsch, he puts Christianity atop and unique among the world religions: "The Christian faith alone dares to maintain revelation in the strict, unconditional sense of the word, because it alone dares to assert, 'the Word became flesh.'" [48] The atonement, God becoming flesh, is the unique revelatory aspect of Christianity.

So here we see Brunner carrying out the implications of his natural theology: examining revelation in creation in the form of other religions. While primitive, polytheistic, or mystical religions, do contain traces of revelation, ultimately their lack of a universal claim to revelation does not put them in the same category as Christianity.[49] Zoroastrianism, Islam, and Judaism do claim to be based on revelation, but lack the redemptive qualities of Christianity. With Brunner and Troeltsch, we can begin to see the flowering of the implications of natural theology in the area of religion. Other religions can be studied theologically, even if the conclusions are somewhat limited and narrow. By the late 1940's there is an openness to discuss other religions that will blossom into the "pluralistic" movement of the latter half of the 20th century. We have come from "pagans", to "infidels", to "heathen", to religions that are limited in their darkened understanding, to religions that contain divine revelation but are ultimately inferior to Christianity, to religions that are equal in their own context with Christianity (cf. Troeltsch). What remains in this overview of attitudes towards other religions is to explore the second half of the twentieth century, when contextual and cultural changes precipitated a full flowering of pluralistic studies.

Paul Tillich & the Beginnings of Dialogue

We turn next to the person of Paul Tillich, who stands at the forefront of the modern concept of pluralism, chiefly because of his attitude toward dialogue. Rather than simply pronouncing judgment on other religions as neo–orthodoxy had, Tillich attempts to engage other religions, however briefly. Tillich asserts that Christianity claims "the event on which it is based is the center of the history of revelation and salvation." [50] Yet he notes that Christianity cannot "overlook the fact that there are other interpretations of history which make the same claim for another central event." The question then comes, how can we assert that the Christian claim is true? Tillich's answer is that the "claim is an expression of the daring courage of the Christian faith." [51] Using Otto's idea of the holy, Tillich argues that there is an irreducible element of mysticism in every religion. [52] Tillich argues for the history of religions being a major source for theology, and is one of the first to include references to other religions in a systematic theology (though almost exclusively in his third and final volume). But he does not fully act on his open stance toward other religions and actively dialogue with them or include much of their thought in his systematics. For that we must look to another work of Tillich's for further development of his thought on the theology of religions.

In *Christianity and the Encounter of the World Religions*, Tillich elaborates the kind of dialogue he envisions Christian theology needing in the future. Tillich begins by setting out the need for a typology of religion, of which Christianity and Buddhism occupy the opposite poles. But having noted their differences, he then attempts to see how Christianity and Buddhism might encounter each other in differing areas. With regard to *telos* (Nirvana vs. the Kingdom of God) and anthropology (humanity as a sinner responsible for the fall vs. humanity bound to the wheel of life with self–affirmation, blindness, and suffering), and in comparing Buddhist compassion with Christian *agape*, Tillich wonders if these major differences spell the end of the dialogue. [53] It would seem that the two religions have little in common. Tillich, however, seeks to transcend the seeming lack of similarity in his celebrated conclusion:

> Does our analysis demand either a mixture of religions or the victory of one religion, or the end of the religious age altogether? We answer: None of these alternatives! A mixture of religions destroys in each of them the concreteness which gives it its dynamic power. The victory of *one* religion would impose a particular religious answer on all other particular answers. The end of the religious age – one has already spoken of the end of the Christian or the Protestant age – is an impossible concept. The religious principle cannot come to an end. For the question of the ultimate meaning of life cannot be silenced as long as men are men. Religion cannot come to an end, and a particular religion will be lasting to the degree in which it negates itself as a religion. Thus Christianity will be a bearer of the religious answer as long as it breaks through its own particularity.
>
> The way to achieve this is not to relinquish one's religious tradition for the sake of a universal concept which would be nothing but a concept. The way is to penetrate into the depth of one's own religion, in devotion, thought, and action. In the depth of every living religion there is a point at which the religion itself loses its importance, and that to which it points breaks through its particularity, elevating it to spiritual freedom and with it to a vision of the spiritual presence in other expressions of the ultimate meaning of man's existence.
>
> This is what Christianity must see in the present encounter of the world religions. [54]

Given such an affirmation of a spiritual presence in the world's religions, one would expect a fair amount of dialogue with that material. Yet Tillich's dialogue with the actual tenets of Buddhism is fairly elemental in nature, consisting of a few pages comparing the Kingdom of God and Nirvana. [55]

Nonetheless, he has begun a dialogue. Tillich himself offers an explanation as to why his systematics are not more inclusive of other religions in his last lecture, 'The Significance of the History of Religions for the Systematic Theologian':

> I must say that my own Systematic Theology was written before these seminars [with noted pluralist theologian M. Eliade] and had another intention, namely, the apologetic discussion against and with the secular. Its purpose was the discussion of the answering of questions coming from the scientific and philosophical criticism of Christianity. But perhaps we need a longer, more intensive period of interpenetration of systematic theological study and religious historical studies. Under such circumstances the structure of religious thought might develop in connection with another or different fragmentary manifestation of theonomy or of the Religion of the Concrete Spirit. This is my hope for the future of theology. [56]

Tillich apparently would have gone on to include more thought from other religions in a systematic work, which would have been fascinating and groundbreaking. Again, Tillich's concerns were shaped by his context: By the end of his life, world events were changing, and the community of nations and religions was becoming more inter–connected. If he had lived longer, Tillich's implementation of his methodology would have been stirring. However, we are left with his extant works, which show a similarity in methodology to Brunner, though Tillich's ultimate *telos* is something much more approaching Troeltsch and his relativism.

In the early 1960's, there were other developments in the attitudes of Christian theologians toward other religions, notably Karl Rahner's notion of the "anonymous Christian". [57] But in the main, the Christian theological world had come to a place where the acceptance of the notion of some type of general revelation present in other world religions, accessible through some form of theology, whether natural or otherwise, was increasing. Theology of religion had become a legitimate area of study in its own right, one that would expand in the later decades of the twentieth century.

The Pluralism Movement

We have already made extensive comments about the modern pluralism movement in an earlier section, and so we note that the contours of the discussion on pluralism which has sprung up in the last 20 years or so have been dominated by the three–fold categories which we have examined earlier:

> 1. the 'conservative' *exclusivist* approach, which finds salvation only in Christ and little, if any, value elsewhere ...
>
> 2. the 'liberal' *inclusivist* attitude, which recognizes the salvific richness of other faiths but then views this richness as the result of Christ's redemptive work and as having to be fulfilled in Christ ...
>
> 3. a *pluralist* position – a move away from insistence on the superiority or finality of Christ and Christianity toward a recognition of the independent validity of other ways.[58]

Each category has a number of adherents [59]: *Exclusivism* – Max Warren, Stephen Neill, Kenneth Crag, Lesslie Newbigin; *Inclusivism* – Gavin D'Costa, Clark Pinnock, Wolfhart Pannenberg, Hans Küng; and *Pluralism* – Paul Knitter, Gordon Kaufman, John Hick, Wilfred Cantwell Smith, Raimundo Panikkar. Thus we have a veritable explosion of literature around the topic.

But before we proceed too much further, a few comments on this three–fold typology are warranted:

1. There is a strong emphasis on Christology – this would seem to point the discussion into a more "exclusivist" direction, or at least cast an exclusivist hue, but in fact it hasn't. For example, Jurgen Moltmann's Christology is very central to his theology of hope, but notice how inclusivist his Christology is: "Outside Christ no salvation. Christ has come and was sacrificed for the reconciliation of the whole world. No one is excluded." [60] Stanley Samartha, Panikkar, Seiichi Yagi, and Aloysius Pieris use Christology to advocate for their "pluralist" perspective. [61]

2. The discussions have evolved from centering around the notion of revelation and natural theology – i.e. how there can be truth in other religions – to discussions centering around soteriology – how and whether someone can be saved in other religions. [62]

3. Another development has been the evolution from the topic of revelation to the topic of the nature of dialogue, with little or no theology actually examined. That is to say, the discussion has turned from theological "content" to that of theological "method." One common characteristic we see is that most exclusivists and inclusivists are content with a discourse on method – how their theology of religion works. Hardly any take the next step, begun by Brunner, and begin to analyze the other religions and either dismiss away (exclusivist), enfold into a Christian system (inclusivist), or synthesize with

(pluralist) Christianity. [63] Those that do interact with other religions tend to fall into the pluralist category, which is not surprising.
4. Very little pluralistic theology is actually taking place, which is surprising given the open and accommodating stances today towards other religions. The direction Tillich was hoping for in systematic theology has not materialized. In summary, there has been very little actual advance in the use of insights from other religions through the many centuries of the history of theology. Despite changing rhetoric and seemingly greater openness, even our pluralistic context has yet to produce a flowering of pluralistic theology.

Three theologians who are exemplary in the pluralist category can be seen as representative of the extent to which actual pluralistic theology is being done. In John Robinson's *Truth is Two–Eyed* (1979), he argues that rather than the typical emphasis of Hinduism being fulfilled in its encounter with Christ, we should see Christianity being fulfilled in its encounter with Hinduism. Robinson shows that the Christian vision is enriched and fulfilled in the light of Hindu perspectives of the personal and material world. [64] Similar to Robinson is John Cobb, who also argues that Christianity can be transformed, this time by Buddhism, in his *Beyond Dialogue: Towards a Mutual Transformation of Christianity and Buddhism* (1982). Four areas in which Christianity could be transformed are: 1) Faith without attachment; 2) The dissolution and transcendence of personal existence; 3) God and emptiness; 4) Pure immanence and partial transcendence. [65] Cobb goes as far as to suggest that Buddhism could experience its own transformation in light of its encounter with Christianity – e.g. that "Amida is Christ", or expressive of the "logos which Christians know as incarnate in Jesus". [66] Raimando Panikkar has two books that would fit this pluralist category – *The Unknown Christ of Hinduism* (1964), and the massive *Myth, Faith and Hermeneutics* (1979). In the *Unknown Christ of Hinduism*, Panikkar argues for a thesis of the unknown Christ, which is that "whether or not we believe in God or Gods, there is something in every human being that does not alienate Man but rather allows Man to reach fullness of being." [67] Panikkar then explores this principle in a detailed investigation of Hinduism. In *Myth, Faith and Hermeneutics*, Panikkar explores hermeneutics and the role of myth and faith, using both Hinduism and Christianity (even Buddhism) as sources. Hinduism is a full partner for Panikkar's investigation, as he is the kind of example we are looking for in our search for a concrete application of the implications of a full–fledged view of general revelation.

But even in these works, the emphasis is on encounter, dialogue – i.e. "what can we learn from each other?", rather than the enfolding or synthesizing of actual doctrines and beliefs into a Christian doctrine. Even if one accepts the "pluralism" (using Hick and Knitter's categories) methodology of developing a "world religion", at best what we see here are discussions of method – e.g. how we might learn from each other, or how do different traditions learn? These writers are pluralists in their theological discussions of method, but they have yet to move towards any "pluralizing" in the major areas of systematic theology – God, creation, humanity, etc.

One area that has been addressed is eschatology, by both Mark Heim's *The Depth of the Riches* (2001), and John Hick's *Death and Eternal Life* (1976). Smart and Konstantine's *Christian Systematic Theology in World Context* (1991) is exemplary, too, of a pluralistic systematic theology. These are the three works we will examine as examples of incorporation. But overall, what we see in the present state of pluralistic studies are a number of works centering around encounter and dialogue, but relatively few theological works addressing major systematic themes. We will end this overview of a theology of religions within the history of Christian doctrine by turning to the three exceptions noted above, exceptions which have yet to impart a serious challenge to Christian theology's inability to take seriously the "revelatory" insight of the other world religions.

Three Examples of Incorporation

John Hick's *Death and Eternal Life*

As the person at the forefront of the pluralist movement, Hick has developed a methodology of pluralism that has been crucial for pluralism studies. Here we will not focus on his theology of religion, but rather the *application* of his theology of religion in a work on eschatology. In beginning his work, Hick states his method, one principle of which is "openness to all data":

> Any writer on death and immortality is likely to start from within a particular religious tradition, which he either affirms or opposes, and within a particular historical culture. But if he restricts himself to the resources of his own religio–cultural milieu he will be interpreting the phenomenon of death from a restricted standpoint which is not warranted by the universality of the phenomenon itself. The experiences and reflections of all human faiths and cultures are relevant to the

> understanding of our mortality. This is par excellence a topic to which a trans–confessional and trans–cultural approach is appropriate… The present enquiry is undertaken from a Christian standpoint. It is not however offered primarily as a contribution to Christian theology, but rather as a Christian contribution to global or human theology. [68]

Hick states his presupposition, that there can be one global theology (not a global religion), only if one gives up a "Ptolemaic worldview" (i.e. from one's own religious viewpoint – cf. the "earth"), and adopts a "Copernican" view (i.e. – religions as different responses to the same Ultimate Reality – cf. the "universe"). [69] After examining contemporary attitudes toward death, Hick analyzes Christian approaches to the after–life, before discussing western and eastern pareschatologies, including survival of the disembodied mind, resurrection of the person, reincarnation, and the Buddhist conception of rebirth. Hick is clearly attempting to bring together thoughts from many varied religious traditions.

In developing his pareschatology, Hick considers the state of the human ego:

> These considerations suggest the likelihood that if the human ego is immortal, or if it persists for a period which multiplies very many times the length of an earthly life, in post–mortem existence occurs in successive sections rather than as one continuous unit. Thus it may be that period death (like periodic sleep) divides up an existence which, as finite creatures, we can only live in limited phases. Its recognition of this is, as we have seen, a virtue of the indian conception of reincarnation. The basic idea does not however have to take the form of repeated rebirths in this world. There is also the possibility of repeated 'rebecomings' in other spheres beyond this world. Instead of a horizontal line, the image is then that of a line moving upwards in what we can call vertical rather than horizontal reincarnation. [70]

Here Hick is appropriating Hindu themes of reincarnation for his theology. Hick goes on to describe the period immediately after death: "The immediately post–mortem phase is subjective and dream–like, and …it can take either a sharply defined form, reflecting an imagination effectively conditioned by a powerful religious culture, or be experienced as a continuation of earthly life." [71] How Hick arrives at this phase is instructive for us. Upon describing the traditional judgment seat of Christ after death, Hick writes, 'but a unitive christian culture, strong enough to impose its distinctive form upon men's after–death experiences, is now a thing of the past, at any rate in most nominally christian countries." [72] So Hick summarily drops this Christian understanding of the afterlife from his scheme. One wonders why, given the presence of a similarly secular presence in Hindu countries and theology, reincarnation is retained as a part of Hick's outline? It would seem that Hick is picking and choosing based on his own proclivities, which is precisely the difficulty one faces in developing a "global theology": it is difficult, if not impossible to be objective, or "copernican", to use Hick's term, and dispassionately neutral. [73] This is a more general critique of postmodernity, and it becomes evident as a major hurdle for all pluralists. A second significant critique of pluralism comes into play here as well, namely that, in developing a global theology, much of what made the religions distinctive is lost, so that what is left is not something that resembles any of the religions at all (which is precisely what Tillich argues for), but a vague generalization. What is left is the type of overarching modern metanarrative whose hegemony pluralism was purporting to get out from under. Describing the life in the next world after this post–mortem phase, Hick relates, "We cannot say anything about the next world beyond the *bardo* state except that it will be a reap spatio–temporal environment, functioning in accordance with its own laws, within which there will be real personal life – a world with its own concrete character, its own history, its own absorbing and urgent concerns, its own crises, perils, achievements, sacrifices, and its own terminus giving shape and meaning to existence within it." [74] Here is perhaps a third critique: Hick seems to be engaging in sheer speculation, a difficulty typical of those not working from within a religious tradition. He has no texts, no oral histories, no theologies to guide him, only imagination. And the result is a generalization that does not approximate Christian theology.

Hick then moves to describe the ultimate state itself, beyond the next stage of human existence. Acknowledging the great difference in the eschatological visions of east and west, Hick argues,

> That the eschatologies developed within Christianity, Hinduism and Buddhism are each essentially open–ended. They are pointers beyond the known which do not profess to delimit the boundaries or describe the contents of that towards which they point. And if we presume that the teachings of these great religious traditions have arisen out of permanently significant experiences at the interface between the human and the divine we shall be open to the possibility that their eschatologies offer convergent indications, each pointing beyond our present human experience and yet each pointing in the same direction. [75]

Using the Christian doctrine of Trinity to explain the eschatological community, Hick speculates, "there will be many persons, in the sense of many centers of personal relationship, not however existing over against one another as separate atomic individuals but rather within one another in the mutual co-inherence or interpermeation (*perichoresis, circuminsessio)* which has been predicated of the Persons of the Trinity." [76] Hick continues, "What Christians call the Mystical Body of Christ within the life of God, and Hindus the universal Atman which we all are, and Mahayana Buddhists the self–transcending unity in the Dharma Body of the Buddha, consists of the wholeness of ultimately perfected humanity beyond the existence of separate egos." [77] Hick thus enjoins the possibility of all the major religions coexisting together in the eschaton. While this arrangement may seem to bring all of the traditions together, this conception, Hick's use of Trinitarian epistemology notwithstanding, has left both Christian and Buddhist theologies behind. Christianity's eschatology is predicated on resurrection (cf. I Cor. 15), both of Christ, and of the believer. This dissolved state does not do justice either to the Buddhist concept of nirvana, which at its core, is nothingness – not unity with Buddha, not perfection, but rather nothingness. Hick has succeeded in melting all the world religions together and coming up with something totally different than all of them, understandable if one is content to "display the pointing fingers within each tradition", [78] as Hick terms his work, but unacceptable if one is developing a grounded, pluralistic theology.

Nevertheless, we do have in Hick an example of someone who has addressed a topic in systematic theology using material from other religions, and for that Hick is to be lauded. So too, is his acknowledgement of beginning from a "christian" (Christian) perspective, even if in the end he lands near none of the religions. In moving beyond a theology of religion toward a topic that actually invites thought from other religions, Hick is again an innovator, though his work is fraught with the difficulties that attend most purely pluralistic endeavors.

S. Mark Heim's *The Depth of the Riches*

Mark Heim, who has written extensively on issues relating to pluralism, provides a counterpoint to Hick's eschatology in *The Depth of the Riches: A Trinitarian Theology of Religious Ends* (2001). Heim states his methodology in his introduction:

> I am convinced that the fruitful and viable path for interfaith relations lies in each tradition developing, from its own particularistic grounds, frameworks for the fullest legitimate recognition of the *distinctive* qualities (those that finally resist assimilation) of the *positive* religious aims of other faiths. This path differs from the

> abstract view of many pluralistic theologies, which wishes to maintain that what is truly religiously important is either an indeterminate point of convergence beyond the particularities of all faiths or some analytic, generic essence of their current functions in human life. It also rejects the convictions (shared by the most pluralistic theologians and the most exclusivistic ones) that there is one and only one actual religious fulfillment to offer to humanity.... [79]

Already distinguishing himself from Hick's exclusively pluralist viewpoint, Heim argues that the grammar of true religious diversity "should also allow for the attainment of religious fulfillments other than the one a given tradition holds to be supreme." [80] Heim starts his theological enquiry from a particular tradition, recognizing that some religious elements are incapable of assimilation, yet he also attempts to do justice to the full diversity of religious thought on eschatology. This is in contrast to Hick's methodology, where Hick begins with a pluralistic epistemology (cf. Heim's description: "an indeterminate point of convergence beyond the particularities of all faiths") and assimilating differing religions' eschatologies, before bringing them together into one system. The contrasts become even clearer as we move forward into Heim's argument.

Heim's innovation is that he holds inclusivism and pluralism together at the same time. He adheres to the primacy of the Christian faith, while affirming that other religious ends, argued for and sought after in other religions, are no less real:

> This book affirms the legitimacy of Christian confession of Christ as the one decisive savior of the world. But it does so by means that will no doubt seem unusual and perhaps paradoxical to many Christians: affirming that other religious traditions truthfully hold out religious ends which their adherents might realize as alternatives to communion with God in Christ. These are not salvation, the end Christians long for. But they are real. [81]

In other words, each religion's ends are to be seen as exactly what each religion is looking for, and by consequence, producing what is intended in its adherents. These ends can be recognized and affirmed from the (different) Christian position. Heim quotes Joseph DiNoia in stating that "Christian theology can affirm 'the distinctiveness of the aims fostered by other religions without prejudice to an affirmation of the unique valuation of the Christian community or of its doctrines about salvation.'" [82] Heim uses Dante's *Divine Comedy* as an example of a variety of human ends concurrently possible, each with its own internal coherence. [83] Heim goes on to place these ends in subordination to the Christian hope: "On the other hand, the true alternative character of these ends allows and even requires a judgment from the Christian perspective that subordinates them to salvation. To realize something other than communion with the triune God and with other creatures through Christ, in the continuing relationship of created being, is to achieve a lesser good." [84]

For Heim, religious ends, what we might call eschatology, are defined as salvation. [85] What Christianity is looking for, argues Heim, ultimately, is salvation, which he then describes broadly as communion with the triune God:

> Distinctive religious ends sought and realized in other religious traditions are grounded in apprehension of and connection with specific dimensions of the divine life of the triune God. They become separate ends by virtue of isolation and limitation, but this does not compromise their reality. Because the nature of God is a communion of the three divine persons, that nature itself has a variety of dimensions. Salvation, the religious end Christians know and seek, is a relational state and not a simple one. It is this composite character that opens the possibility of different but real religious ends, ones whose limitation leads to distinct definition, whose isolation leads to a special purity. [86]

Key for Heim is the role of the Trinity, which, in reviewing both Panikkar and Smart, he concludes "provides a matrix within which Christians can recognize distinctively different religious practice as relation with God." [87] Because of the primacy of relationship within the Trinity, relationship (to and with God, humanity, and the rest of creation) is key to Heim's definition of salvation. This relationality is then played out in the differing religious responses to God's entreaties: "The situation I have described is one where God's universal salvific will is the co–determiner of human religious ends. Because salvation is a free relation, God has taken absolute control only over God's end of the relation. Religious ends are co–determined by persons themselves in their relational activity." [88]

Heim brings us closer to a postmodern eschatology informed by other religions, but a major critique of his work is that he does not actually quote other religions. His work is more or less a treatise on how his thesis – that other religions can fulfill their religious ends, alongside of Christianity's – can work epistemologically via the Trinity. He never delves into the specifics of the other religions. For example, if one accepts Heim's thesis, how exactly do reincarnation (Hinduism), complete emptiness (Buddhism), and resurrected life with God (Christianity), all become operative at the same time for people with different belief systems? While advocating his thesis, and showing that it would be possible from within the framework of Christian theology, Heim never shows us how it actually would work. And, as for our concern in this book, Heim remains solidly in the camp of those who write about a theology of religion from an inclusivist point of view (cf. methodology), but never get around to developing a Christian theology that would include those "less–developed" religious viewpoints.

In the end, Heim has not put forth a full–orbed Christian eschatology (or more to the point of his book, soteriology) that incorporates or synthesizes material from other religions. His book does, however, provide an epistemological framework for how that might happen. In this regard, we are left, as with Tillich, standing at the door. It will remain for others to draw out the detailed, real–life theological implications of Heim's suggestions. [89]

Smart & Konstantine's *Christian Systematic Theology in a World Context*

We are re–visiting Ninian Smart simply because of the paucity of sources in this area of pluralistic systematic theology. While we have already described Smart's methodology and contributions in the area of the doctrine of creation, it now falls to us to examine how he uses sources from outside the Christian tradition. In their 1991 book, *Christian Systematic Theology in a World Context*, Ninian Smart and Steven Konstantine present a systematic theology that is purposely inclusive of other traditions besides Christianity, while still retaining a Christian point of view. Smart sums up his position as "neotranscendentalist, pluralistic, social Trinitarian, universalist panentheism embedded in a soft epistemology." [90] Smart wants Christian theology to include the language, concepts, and cultural forces of the world, including religious and secular worldviews.

After a lengthy methodological discourse (something all too typical as we have seen in others, as well), Smart moves to elaborating his systematic theology via the epistemology of the Trinity. Smart holds to a social Trinitarian view, and stresses the individuality of the three persons of the Trinity, and the way those three persons commune as being important for understanding plurality. [91] This is his starting point. However, it is important to note that even here, Smart refers to both Buddhist sources and Hindu sources in corroborating his theory about the Trinity. [92] He then goes on to explain the Trinity's threefold work in the world – "divine creativity," "divine embodiment," and "spirit and psyche." In his analysis, we see Smart's openness to other religions, even other secular thought processes. This is based, as we have previously seen, in his panentheism.

In both his chapters on the "divine embodiment" and "spirit and psyche", Smart makes ample use of both Hindu (*avatāra*), and Buddhist (*bodhisattvas*) symbols. It is important to see that these references are not merely used as foils, or fleeting comparisons; Smart includes their ideas in his Christology. He writes, "the plural character of the religions helps them to criticize and correct each other: to keep each other honest." [93] In his pneumatology, Smart makes frequent references to psychological and biological sources as well.

Obviously, Smart is open to a wide variety of sources for his systematic theology. Two points about his approach are laudable. The first is his avowedly Christian beginning point. While he is open to other sources, Smart is clear to begin his theology from a Christian point of view. And, in the main, though he cites other religions and sources, his work can be broadly described as a Christian theological one. This approach, though it is fraught with the difficulties associated with a purely pluralistic theological endeavor, is pointing us in the correct direction for pluralistic studies. Secondly, as regards the specific focus of this section, Smart uses extensive imagery and theological data from other religions in crafting his positions, though that material from other religions is quoted more often in his thoughts on methodology and theories of religions than it is in his purely constructive chapters on the Trinity. Clearly Smart is attempting to involve other traditions in his theology. Along with Hick, he is our only other example of that process.

Where I would be critical of Smart is in how he uses Christianity as a starting point for his theology. He makes it clear that he prefers the term "Christianities" as opposed to "Christianity" as a moniker, emphasizing that there has not been one tradition. While the varieties of Christendom are well known to theologians, the general character of orthodoxy has exhibited an equally heavy weight on Christian theology. Orthodoxy forces us to think in terms of "Christianity". And this is not a negative, because orthodoxy reminds us that where theology has strayed it ceases to be called truly Christian. Of course, the Manicheists and Donatists would be classified under the Christian tent in comparison to other world religions, but the Christian tradition has relegated them to the sideline as heterodoxy. The practical import of Smart's use of "Christianities" is that his theology often moves outside the orthodox realm (for instance, one cannot claim that the universe is God's body (cf. p. 209) and continue to argue for a creator/creature distinction, something crucial for a Christian doctrine of creation). He begins with and is guided by Christian categories, but ends with something that, while it is informed by other traditions, ends up being not rooted in the Christian tradition. Smart is attempting to be both pluralist and self–consciously Christian in his theology, a difficult balancing act. Nevertheless, Smart has blazed a trail in the area of systematic theology, particularly in the area of creation, which can serve as an example for us.

Conclusions

What have we learned from our foray into how Christian theologians have used material from other religions?

1) There has been a real development in Christian theology in its stance towards other religions. From being concerned about paganism creeping into the church, to seeing that other religions had potentially something positive to contribute to humanity, to a few arguing for outright equality with Christianity – the role of other religions in the field of Christian theology has increased dramatically. In the first millennium of the church, the purity of the church was at stake, as even communicating with a believer in another religion was suspect. As we finish the second millennium of the church, the evolving world situation, with differing cultures and beliefs coming into contact with one another as never before, has created a new set of questions, leading to the development of theologies of religion. There are now several different typologies available to classify how one might discuss other religions. Correspondingly, attitudes toward other religions have evolved as well, from hostility to tolerance to dialogue. Now we are on the verge of the incorporation of insight from other religions into our theologies.

2) The theological rationale for interacting with other religions has evolved as well. Theology of religion has come upon the scene as a category for theology only recently, but early theologians were nevertheless offering opinions on other religions. We can note a progression of openness to revelation being present in other religions, but at all times theologians were reflecting the concerns of their day. From the beginning of the church, the early theologians were open to philosophical influences, and were even able to acknowledge the possibility of revelation in other religions. As we move to Aquinas, we begin to see the possibility of learning about God without appealing to

special revelation, and the development of a "natural" theology. The Reformers were concerned to develop the Protestant church against the backdrop of Catholicism, and were not thinking in terms of developing outside sources of revelation. As the centuries pass, Schleiermacher and Troeltsch offer for us a new openness to other religions, by accepting that revelation is present in other religions, and so we see a move away from exclusivity toward inclusivism, even relativism in Troeltsch. We begin to note the rise of theology of religion as a field of study in its own right. Schleiermacher sees the rationale for studying other religions to be the innate essence of religion that all humans have access to. Barth sided with the Reformers in affirming a "no" to natural theology, while Brunner, in his work on revelation, provided us with a rationale for seeing "truth" in other religions. Ultimately, Brunner, along with many modern inclusivists, puts Christianity atop the development of world religions, with a unique salvific role. Once we enter the late 20th century, however, discussions about natural theology and the role of revelation have been left behind. There has not seemed to be as pressing of a need to develop a basis for exploring general revelation. It is assumed that there is some revelation in other religions; the question is, how much? [94] There is a need for a re–visited, robust theological rationale for interacting with other religions. And, as we indicated in our section on natural theology, perhaps the pivotal axis should be pneumatology.

3) In admitting that most contemporary theologians who write on this matter are inclusivists or even pluralists, we might expect to see a corresponding rise in interaction with material in other religions. But this is not the case. Most contemporary theologians are content with second order reflection, commenting on other religions via a theology of religions, or through the theme of dialogue, with little engagement with the actual sacred texts of Islam, Buddhism, or Hinduism. This is even true of theologians in the later half of the 20th century. Our three recent exemplars of those who might interact with the theologies of other religions receive mixed reviews at best. Hick's attempt, while laudable, lands him in the hot water that usually accompanies such purely pluralistic endeavors: His scheme has no relation to any particular tradition – it is above and beyond them. And, as such, his scheme does not do justice to the details of any tradition. It is mostly speculation. Heim's approach seems more promising, beginning with an avowedly Christian starting place, but in the end, his work is merely another theology of religion. Though he puts in place an epistemology of how an eschatology

informed by other religions might look, in the end he doesn't really develop one. His interaction with other religions remains quite limited. Smart starts and works within a Christian framework, and he helpfully includes material from other religions. Ultimately, however, he succumbs to the lure of pluralism, ending with a work that is not Christian enough for orthodoxy, and vague enough to not be tethered to any tradition. Moreover, aside from Hick and Smart, hardly any major systematic theological works make use of material from other religions. This is surprising, given the general openness of many theologians towards revelation in other religions. We are convinced that theologians are sitting on an untapped resource.

What remains then is for a constructive theology that is both inclusive enough to include material from other religions, and sure enough of its Christian moorings to be presented to the theological community as Christian theology. It has been difficult to do both at the same time. However, the Christian tradition has developed theological resources – natural theology and general revelation are two of the more familiar ones – that make it possible to do so. Our highlighting of the cardinal role of the person and work of the Holy Spirit in this endeavor displays another vehicle for being both open to the world and committed to a tradition. Providentially, we are in a day and age – that of postmodernity – that asks, even requires, us to construct our own theology while being a part of the worldwide theological community. To remain true to our tradition, while at the same time be cognizant of the world around us, and even learn from it, is the challenge of our generation of theologians.

CHAPTER FIVE

A Model From Old Testament Wisdom Literature

Aside from the theological rationale present in natural theology, Christian theology has another place to turn to develop a methodology for using material from other religions: we have a model within scripture itself. This is ironic because that is the last place exclusivists would expect to find an example of how to incorporate other religious thought into the Christian tradition. But nevertheless it is there in the wisdom literature of the Old Testament. It will prove helpful, then, to elucidate this alternate picture of "incorporation." Once we have explored this model, we will then be in a position to return to our doctrine of creation to expand it through interaction with other religious viewpoints.

Any foray into Ancient Near Eastern wisdom literature is met with a stunning variety of sources. From the sages of Egypt, to the lists and literature of Mesopotamia, to the tradition of ancient Israel, the vast body of work known as wisdom literature stretches both the heart and mind. Our particular interest in this subject matter stems from this great plurality of perspective. Wisdom literature's cosmopolitan nature makes it a compelling study for anyone interested in pluralism. Wisdom literature is also linked theologically to our main area of study, that is, a doctrine of creation.[1] So combining these areas – a doctrine of creation and pluralistic study – creates a tantalizing intersection in wisdom literature that can be a model for us in doing pluralistic theology.

Our inquiry is focused: The main goal is to analyze how Israel interfaced with her neighbors in wisdom literature, neighbors with a vastly different theology, cosmology and anthropology. Certainly there are similarities between the different worldviews, and that will be important to note as well.

But there are differences – even significant ones. These differences somehow found their way into Israel's sacred literature in the form of incorporated religious writing. What we hope to find is a key to how Israel both preserved her own unique viewpoint, while at the same time assimilating other belief system's points of view. As we have been discussing, this increasingly pluralistic world is suggesting to us that we must find a way within Christian theology not to hibernate within our tradition, or assimilate with other traditions, but rather to accentuate the key points of our narrative and remain open to adapting, enhancing, or even correcting our narrative with other stories. We hope to find a model within Christianity's own scriptural tradition that will better enable us both to nurture our own perspective while at the same time learning from others.

This chapter will be divided into four sections. In the first we will make some introductory remarks about wisdom literature in general, paving the way for our second section, which will examine the nature of the Old Testament's wisdom literature's dependence on different religious sources. The third section will see us look at four different case studies, focusing on the specifics of comparing Israelite and other Near Eastern literature. In our final section, we will draw some conclusions as to the theological emphases that guided Israel in crafting her wisdom literature in such a multifarious environment.

Wisdom Literature

Though Preuss and others have stressed wisdom's differences with the rest of the canon, [2] our starting point is that the theological center holding together a Biblical theology made up of the entire Old Testament is God. [3] Zimmerli, in a defining article for subsequent wisdom scholarship, noted that wisdom "thinks resolutely within the framework of a theology of creation". [4] Zimmerli argues that Wisdom is connected with the rest of Old Testament theology through the admonition given to humanity in Gen. 1:28 to "have dominion" over the earth.[5] This dominion is seen in that humanity is not passive, but rather active, seeking its place in the created world. This activity, this independence, this impulse to master the world is authorized, even blessed by God. It is this spirit that imbues wisdom literature.

While Zimmerli's emphasis is on the anthropological thrust of wisdom,[6] the real crux of the link between wisdom literature and the rest of the Old

Testament is not the concept of creation but the presence of the Creator. Yahweh is both Creator and Redeemer, holding law and wisdom together, providing the connection within his own being. The link is theological in its purest sense. Zimmerli himself seems to argue for this, citing "the fear of the Lord is the beginning of knowledge" (Prov. 1:7), and positing that wisdom's theology of creation is always subordinate to God's will.[7] Furthermore, Zimmerli points out that Ecclesiastes knows Genesis 1, and encounters a Creator behind everything; in Ecclesiastes we see the freedom of God who acts and never reacts. [8] Even Zimmerli, with his stress on anthropology, nevertheless recognizes the theological hinge that is at the core of placing Old Testament wisdom within the overall Old Testament canon – God himself.

Secondly, it is important for us to recognize a major sub–theme that brews under the surface of wisdom literature, one that has a direct link to the doctrine of creation – the role of societal chaos and the situation of a dwindling society (in Israel particularly). James L. Crenshaw suggests that Ancient Near Eastern wisdom literature developed in a context of continuing disorder and chaos, forcing thinkers and theologians to question the concept of divine order, particularly in Job and Qoheleth. This tendency was not limited to Israel, with its strong monotheism, but encompassed the other, more polytheistic and henotheistic religions of the area. Crenshaw writes,

> Now the order established in the beginning by the creator is ever subject to the threatening forces of chaos, both human and divine. The eruption of disorder came at the very point where the retribution scheme gave way; this crisis of confidence in divine power or goodness surfaced early in Mesopotamia, Egypt, and Israel. The literary deposit of this heroic struggle is living testimony to the awesome power of the forces of chaos that placed a question mark over the belief in a correspondence between virtue and reward. [9]

Crenshaw sees theodicy, then, as a major subtheme of wisdom literature, as humans protested against what they perceived to be unjust suffering. As society disintegrated, and traditional answers (in this case, a tight cause and effect matrix) failed to explain adequately the situation, people began to look for new answers, and they found them in the form of wisdom, an explanation for life that mitigated life's capriciousness, acknowledged it, and gave some reasons to persevere in the midst of it. The answers in wisdom literature are as varied as wisdom literature itself, from the reaffirming God of wisdom behind Proverbs, to the overpowering God of Job, to the sharp skepticism of Ecclesiastes. [10] Crenshaw argues that it is creation that is used explicitly in this context of chaos as the trump card to provide wholeness and security. [11]

We will be arguing that one of the ways Israelite wisdom literature attempts to gain a sense of order is by adapting the Egyptian concept of *ma'at*. Ernest Würthwein summarizes the Egyptian concept of *ma'at* or "order":

1. Life proceeds according to a fixed order.
2. This order is teachable and learnable.
3. Man is thereby handed an instrument with which to determine and secure his way through life. Because,
4. God himself must pattern himself according to this order, this law. [12]

We see something similar to this particularly in the book of Proverbs; but is the Israelite understanding of "order" influenced by *ma'at*? There is no doubt that there is some evidence of borrowing, but even then, Israel did not leave this concept "as is". She augmented it; but to ask how and why is to get ahead of ourselves. We will return to this point later in the paper as we examine *Amenemopet*. [13]

This mixture of a desire for order, a strong theocentrism, and a burgeoning sense of critiquing that order, are all combined into the paradoxical thrust that is at the core of wisdom literature. We must now turn to the issue of how and if this literature was influenced by other ancient near eastern origins.

Dependency?

Our task here is to examine in general the possibility of Israel borrowing from her neighbors for her wisdom literature. And in so doing we hope to discover a model for theological incorporation that we can use in our present–day pluralistic, postmodern era. It is clear that wisdom was an international phenomenon, encompassing the entire ancient Near East. Both Egypt and Mesopotamia had a thriving wisdom tradition, of which we have many extant sources. Within the Bible we have reference to a yet–to–be–discovered Edomite wisdom tradition (Jer. 49:7). There is also a comparison of sorts, an ancient pluralism, in I Kings 4:30, where Solomon's wisdom is said to be greater than that of the entire East and of Egypt. This comparison is remarkable in that one can hardly imagine a similar statement placing Yahweh's prophets, or Moses' law, for example, in comparison with their surroundings. [14] Further, Proverbs 30:1 and Proverbs 31:1 claim respectively to be imparting "The words of Agur son of Jakeh of Massa", and "The words of Lemuel, king of Massa, which his mother taught him", and neither of these two figures are Israelites. [15] So that it is clear that Israel was, at minimum, aware of and interacting with her surroundings in the realm of wisdom. We will briefly describe each region's literature and some general affinities with Israel's writings.

Egypt's wisdom literature can be divided into two broad categories – Instructions (*Sebayit*) from sages (e.g. *Ptah–hotep, Ani, Amenemopet*); and other Discussion Literature (e.g. *Papyrus Lansing, Dispute over Suicide, Lamentations of Khakheperre–sonbe*). The *Sebayit* have an emphasis on training courtiers, with instructions on statesmanship, correct speech, who to trust, fate and responsibility, etc. The other Discussion Literature questions life's meaning, and examines different alternatives (perhaps protests?) to the *ma'at*–influenced literature of the *sebayit*. For our purposes, it may be said that in general many of the instructions are based on tradition (even *ma'at*). "The teacher transmits to his students the intellectual tradition that has accumulated over the years." [16] Thus, a "fundamental conservatism" pervades these instructions. The Old Testament does not have a category of writings that corresponds directly to this, though Proverbs may be a parallel in its development and transmission through families and clans.

However, some of the religious/moral content imbedded within the writings does correspond with biblical issues. For example, *Amenemopet*'s belief that right conduct stemming from a love for God finds direct resonance with the Old Testament. On the other hand, the royal instructions to kill those in the court that talk and the observation that one's fate cannot be altered in *Merikare*, along with the admonition to not trust anyone in *Amenemhet*, are not to be found within the biblical corpus. The *Papyrus Insinger* contrasts the wise man and the fool, and puts forth an equilibrium that sin can distort, prompting God to punish acts that distort the balance (cf. Proverbs). *Songs of the Harper* explore skepticism and hedonism, while *Dispute over Suicide* (or the *Man Who Was Tired of Life*) has many parallels with Job. The *Lamentations of Khakheperre–sonbe* exhibit a pessimism that is eerily similar to Ecclesiastes. While there are certainly wide differences in some of the specifics between Egypt and Israel, there is enough common ground to posit a collective body of literature.

In Mesopotamia, Sumerian, Babylonian, and Assyrian literature flourished alongside of Israelite wisdom. [17] There are Lists (*Onomastica*), Instructions (e.g. *Instructions of Suruppak*), Proverbs, and other writings. It is in these other writings that we encounter the most striking resemblances to biblical literature, though there are similarities with biblical themes throughout the region. The *Epic of Gilgamesh*, while not wisdom literature in the classical sense, contains a number of wisdom themes about the meaning of life and the quest for immortality. *I Will Praise the Lord of Wisdom* has similarities with Job, as does *The Babylonian Theodicy*, which is a dialogue about human misery. The *Dialogue of Pessimism*, a discussion between a master and a slave about the meaning (lessness) of actions has been compared to Ecclesiastes. The Aramaic *Words of Ahiqar* include wisdom sayings that are similar to the book of Proverbs. Again, we see a number of similar themes both in Mesopotamia and in Israel.

These few, brief examples are enough to show us that there were a number of concurrent works being developed throughout the ancient Near East. We see common themes emerging, ones that have similarities with Proverbs, Job and Ecclesiastes. Many Mesopotamian and Egyptian works press the issue of theodicy – why do the righteous suffer? In the case of Egyptian wisdom literature, this may even be seen as "protest" literature against the prevailing *ma'at* perspective. Mesopotamian literature is much stronger on this point (e.g. *I Will Praise the Lord of Wisdom*, and the *Babylonian Theodicy*), however none of these are as outspoken as Job. [18] Day also mentions that an Akkadian work comparable to *I Will Praise the Lord of Wisdom* has been found at Ugarit, proving that the theme of the righteous sufferer was known very early in Canaan. [19] Such texts *might* have been available to the author of Job. A further common theme is that of the futility of all human endeavor. Lambert surmises that the *Dialogue on Pessimism*, with its extreme positions, draws on a common wisdom theme found all the way back in the Babylonian edition of *Gilgamesh* (c. 1700 BC). [20] This theme – "vanity of vanities" – continued in Mesopotamia through at least the 7th c. BC; thus Ecclesiastes was re–presenting an old theme in Israelite garb. [21] Of course, we also have the proverb. The similarities between the Aramaic *Words of Ahiqar*, *Proverbs*, and *Amenemopet* are great, with a deed/consequence framework, instructions for the young, and calls for discipline. [22] This provides us with one more example of the commonality present in the ancient Near East.

It seems clear that there were a number of similar themes being developed all around Israel, so that establishing a clear, direct linkage between two texts may not be necessary. Crenshaw has pointed out that "the amazing similarities in the thinking of sages throughout the Fertile Crescent do not demand a hypothesis of borrowing, for most resemblances occur when the teachings address universal problems." [23] Crenshaw's point is undercut however, by the very nature of wisdom literature. The sheer prevalence of common themes and understandings suggests some partaking of the common "soup" of literature present at the time. While specific linkages are difficult to claim, borrowing from a wellspring of thought common to an area is borrowing just the same. We will be arguing that in some cases (i.e. Amenemopet) direct borrowing was happening, and in others (Qoheleth, Job), borrowing from the "soup" was taking place.

This is not to say that all wisdom literature espouses essentially the same themes. There are significant differences in the nuances, the motivations, and the consequences involved in the differing sources. And indeed, the focus of this chapter is on how Israel crafted her own wisdom literature in the face of such an international phenomenon. For while it is clear that Israel promotes similar topics as her neighbors, how she uses similar material borrowed from other cultures will tell us much about her own theology. Ernst Würthwein quotes J. Hempel in writing, "It is part of the essence of OT faith that it 'bears a polemical and usurping character, that it does not remain static, but lives in constant dialogue, takes hold of assimilable thoughts, concepts and ideas out of other religions and, transforming them, incorporates them into itself,' but also clearly rejects what would endanger it." [24] This *transformation* is what we are after in this chapter, not only for its benefits for wisdom study, but, even grander, for a model of how to do pluralistic theology in a postmodern setting.

Case Studies

In examining how Israel borrowed from her neighbors theologically, we must remember that success in our endeavor here does not depend on the establishment of a clear linkage of one text to another, but rather, in light of our argument above (common, general wisdom themes), an analysis of similar ideas and constructs. Nevertheless, our first text is one that is perhaps the most celebrated example of borrowing in biblical wisdom literature.

The Instruction of Amenemopet and Proverbs

The Instruction of Amenemopet, from the New Kingdom period of Egypt (c. 1200 bc), has compatibility with various Old Testament passages. For example, the final chapter of *Amenemopet* reads very comparably with the introduction to the book of Proverbs (cf. Prov. 1:1–6),

> See thou these thirty chapters:
> They entertain; they instruct;
> They are the foremost of all books;
> They make the ignorant to know.
> If they are read out before the ignorant,
> Then he will be cleansed by them.
> Fill thyself with them; put them in thy heart,
> And be a man who can interpret them. [25]

In the 4th chapter of *Amenemopet*, we see a contrast between a "heated man", who is "like a tree growing in the open" and headed for destruction, and a "truly silent man", who is "like a tree growing in a garden". That tree "flourishes and doubles its yield." [26] Compare this to the analysis of the righteous and wicked of Psalm 1.

But the most striking similarities are found in Proverbs 17:22 – 24:22. Prov. 22:20 reads, "Have I not written for you thirty sayings of admonition and knowledge, to show you what is right and true, so that you may give a true answer to those who sent you?" (NRSV) The corresponding passage quoted above from *Amenemopet* seems to firmly establish a link between *Amenemopet* and Proverbs. Further similarities may be seen in the following examples:

> Do not associate to thyself the heated man,
> Nor visit him for conversation (*Amenemopet* 11:13–14; *ANE*, 240)
> Make no friendship with a man given to anger,
> Nor go with a wrathful man (Prov. 22:24, RSV)

> Do not carry off the landmark at the boundaries of the arable land
> Nor disturb the position of the measuring–cord;
> Be not greedy after a cubit of land,
> Nor encroach upon the boundaries of a widow. ...
> Guard against encroaching upon the boundaries of the fields,
> Lest a terror carry thee off.
> One satisfies god with the will of the Lord,
> Who determines the boundaries of the arable land.
> (*Amenemopet*, 7:12–15, 8:9–12; *ANE*, 238–239)

Do not remove an ancient landmark or enter the fields of the fatherless;
For their Redeemer is strong;
He will plead their cause against you. (Prov. 23:10–11)

Be not greedy for the property of a poor man,
Nor hunger for his bread.
As for the property of a poor man, it (is) a blocking to the throat
(*Amenemopet*, 14:5–7, *ANE*, 240)
Do not eat the bread of a man who is stingy
Do not desire his delicacies
For he is like one who is inwardly reckoning
(Prov. 23:6–7, RSV; alt.: "like a hair in your throat")

These examples are enough to show a strong correspondence between Proverbs and Amenemopet. Which then came first, Proverbs or Amenemopet? This issue has been debated for decades now, and is nicely summarized in Glendon Bryce's A Legacy of Wisdom (1979). Bryce goes through detailed theories of Semitic Dependence, including those of Budge, Gressmann, Sellin, and Herzog. He also outlines the theories of Egyptian Dependence from Oesterley, Kevin, and Drioton. In the end, Bryce suggests three stages of Semitic development of Egyptian material: Adaptation – few changes; assimilation – important modifications; and integration – little remains of the original contribution. [27] And he argues for Proverbs assimilating Amenemopet in a section concerning the introduction of the two passages, which is illustrative of how Proverbs assimilates Amenemopet elsewhere:

> This comparative study of the introduction of the two collections reveals that the assimilative process involves a creative use of source materials. The blending of the two is so skillfully achieved that only such a careful analysis reveals the relationship between them. The Hebrew ideas have not been fitted into a rigid scheme provided by the Egyptian model, nor do the elements derived from this source appear in a contrived manner. The opening plea appears to be a perfectly natural example of Hebrew parallelism. However, upon closer analysis it proves to be a remarkable fusion of Hebrew parallelism with Egyptian word order. The rest of the introduction moves progressively through the precepts of Amenemopet, picking up the key words yet shaping the whole into a unique introductory appeal. Apparently, the delicate balance between these two was lost when the tridents could no longer understand the development of the passage. Their attempt to integrate it into the Hebrew tradition obliterated its principal referent and apparently corrupted other important elements. [28]

Bryce posits that kingship and the royal court was an important motivating force behind the assimilation of foreign wisdom for Israel.[29] It now seems clear that Proverbs was dependent on Amenemopet. [30]

For our purposes, we are interested in how and why the older material was adapted, assimilated, or integrated. One central concern in Amenemopet is that of the contrast between the "heated" man and the "silent" man, a concern that is found in much Egyptian wisdom literature. [31] The silent man accommodates himself to the divine purpose and is found pleasing to the gods. The heated man is in conflict both with the gods and himself. The similarities here with the Israelite dichotomy of wise/fool, and good/evil are obvious, though Proverbs' dichotomies are imbued with a more religious character (e.g. Prov. 1:7; 3:5–6; and especially 22:19). The motivation for good deeds is "the fear of the Lord", that "your trust may be in the Lord." Now certainly religion is not lacking in Amenemopet (cf. chs. 18, 20, 21), but the references are to the gods in an almost abstract manner.

Much more prevalent is the Egyptian concept of fate – i.e. ma'at – "there is no ignoring Fate and Fortune" (Amenemopet 7:11) and "Do not discover for thy own self the will of god, without (reference to) Fate and Fortune" (Amen. 21:15–16). What is emphasized is that the gods or God is in control (e.g. "God is (always) in his success, whereas man is in his failure" Amen. 19:14–15), though distantly, and humanity's duty is to accept his or her fate in life. This distance calls for acquiesance from the human – "For surely thou knowest not the plans of god, lest thou be ashamed on the morrow. Sit thou down at the hands of the god" (Amen. 22:5–7) – and is also the source of divine mystery: "one thing are the words which men say, another is that which the god does" (Amen. 19:16–17).

The Proverbs passage differs from Amenemopet mainly in theological areas. Proverbs certainly has an anthrocentric thrust, and the emphasis on behavior in Proverbs mirrors that in Amenemopet. However, humanity in Proverbs is not alone, twisting in the wind of his/her accomplishments while the gods/God moves forward unthwarted. Humanity's goal is not to find the secret balance of ethical behavior that will please a distant God or an impersonal "order", but rather to "trust in the Lord" so that the ethical behavior will be intertwined with God's will and reward. The God of Proverbs is much nearer than the gods of Amenemopet, and his ways are not so inscrutable. Moreover, He is not dependent on some impersonal order (ma'at), but is rather the Orderer. To be sure, God is not known in totality in Proverbs, and there is an emphasis on God's transcendence. Nevertheless, humanity's behavior is also linked with an immanent God. Like Amenemopet, the Lord of Proverbs is in control, pleading the cause of the poor and fatherless. This control, however, is more advocative in Proverbs as Yahweh works for the weak, again emphasizing the "personalness" of God (cf. Prov. 23:10–11). In both texts we see that morality is central – the ethical motivation to help others, though, is less centered in love for the neighbor (NT concern) than in doing so simply because that is the way the universe ought to be. Both texts have a telos – Proverbs holds out a "future" and a "hope" for its adherents (cf. Prov. 24:14, 20), while Amenemopet states that "thy body will prosper upon earth" (Amen. 4:2). In the main, the central difference between these two texts hovers around the divine "distance". While the God of Proverbs may not be as close to humanity as the incarnated Jesus of the New Testament, Yahweh is still nearer to, and more involved in, human life than the gods of Amenemopet.

It remains for us to ask ourselves the question of whether a theological analysis of these different texts gets us to the controlling factors in how Israel adapted material from Egypt. It appears the older material was used because 1) it corresponded with material already present in the Israelite tradition; and 2) it provided a platform on which to emphasize distinctive elements of the Israelite tradition. As to whether the Egyptian material provided a challenge to Israelite tradition in the manner of contemporary worldviews – i.e. inclusion of Egyptian material brought out new viewpoints not emphasized in the Israelite tradition – this we do not see. It may be that the adaptation of Amenemopet attempted to highlight precisely the theological differences we have already noted. The very fact that a source outside the Israelite tradition was used certainly suggests insight was gained, or correction was offered; if not, why would Israel bother with the other source? In the final analysis we simply do not have any material to get at the motivations and rationales of the people behind the adaptation other than the theological differences we can note.

The Epic of Gilgamesh & Ecclesiastes

Our second analysis is a lesser–known comparison. For us, comparing The Epic of Gilgamesh and Ecclesiastes will serve as an entré into the broad wisdom literature theme of carpe diem, and the other themes that Ecclesiastes introduces. The Epic of Gilgamesh is not typically seen as wisdom literature, but it certainly contains wisdom themes, and those will be the parts of the story we will focus on. Scholars have long noted the striking parallels between Qoheleth's advice on enjoying life in Eccl. 9:7–9 and the barmaid Šiduri's charge to Gilgamesh. The following illustrates the parallels:

Ecclesiastes	**Gilgamesh**
For the living know that they will die, but the dead know nothing and they have no more reward; but the memory of them is lost. Their love and their hate and their envy have already perished, and they have no more for ever any share in all that is done under the sun. Go eat your bread with enjoyment,	Gilgamesh, whither do you rove? The life you pursue you shall not find. When the gods created mankind, Death for mankind they set aside, Life in their own hands retaining. As for you, Gilgamesh, let your belly be full,
and drink your wine with a merry heart for God has already approved what you do.	Make merry by day and by night. Of each day make a feast of rejoicing, Day and night dance and play!
Let your garments be always white; Let not oil be lacking on your head.	Let your garments be sparkling fresh, Your head be washed; bathe in water. Pay heed to the little one that holds on to your hand.
Enjoy life with the wife whom you love, all the days of your vain life which he has given you under the sun, because that is your portion in life and in your toil at which you toil under the sun.	Let your spouse delight in your bosom! For this is the task of [mankind]

In 1905, Grimme first noted that Ecclesiastes is dependent on Gilgamesh at this point, and Day, Jones, and others have seconded this view. [32] Similar perspectives on life can be found in Egyptian or Greek literature and may have simply been a part of the "soup" of ancient Near Eastern wisdom literature that Qoheleth drew from. [33] In either case, Qoheleth more than likely was dependent on other religious sources, though finding a clear, direct link is not foundational to proceeding to analyze how Qoheleth adapted whatever sources were at his/her disposal.

The Epic of Gilgamesh recounts the failed attempts of the hero Gilgamesh to find immortality, and was probably written by a single Akkadain author c. 2000–1600 bc. Along the way, Gilgamesh struggles with mortality and the limits of human achievement. Gilgamesh declares, "As for mankind, numbered are their days. Whatever they achieve is but the wind" (Old Babylonian version, tablet 3:7–8). Compare that to Eccl. 5:16: "Just as they come, so shall they go; and what gain do they have from toiling for the wind." [34] Another similarity is seen in the image of the threefold cord: "two are better than one … If they fall, one will lift up his fellow … A threefold cord is not quickly broken" (Eccl. 4:9–12). In both *Gilgamesh and the Land of the Living* and the *Epic*, Enkidu reminds Gilgamesh, "Two men will not die; the towed boat will not sink, A towrope of three strands cannot be cut … You help me and I will help you, (and) what of ours can anyone carry off?" (5.2.20–25; though better preserved in *Gilgamesh and the Land of the Living*).[35] These two remarkable parallels are enough to point to a strong connection; Jones closes his article by ultimately suggesting literary dependence based on 1) the existence of five similar details in two texts thousands of years apart, and 2) the fact that these details appear in the same order. [36]

But how does Qoheleth make use of whatever materials were available? What is emphasized? Certainly there are many similarities in the three passages quoted, but as Day has noted, "one thing we may be confident of is that, whatever foreign influences there may have been, Qoheleth had very much made the ideas his own, for he clearly felt things passionately, and was not simply a slavish follower of foreign philosophies." [37] This originality of Qoheleth points us in the direction of his overall religious viewpoint as compared to that in *Gilgamesh*. For Qoheleth, God is unknowable (11:5), and Qoheleth complains that God's activity is a mystery. But, "not for a single moment did [Qoheleth] entertain the possibility of atheism, at least in his written legacy." [38] Is Qoheleth's view of God such that humans can never know what God wants, and so a kind of fatalism in the face of indiscernible determinacy is what is needed? This is the viewpoint we encounter in *Gilgamesh*. Crenshaw has pointed out that Qoheleth "seems to know far more about God than his theology of divine mystery allows." [39] Not only does Qoheleth make assertions about God's will notwithstanding his hiddenness, but he enumerates various gifts that God gives to humans: eating, drinking, and enjoying life – something that surprisingly (or not so surprisingly to an Israelite audience steeped in Genesis) Qoheleth states God has already approved (9:7). In short, the religious view of Qoheleth is much more involved and personal than is that of Gilgamesh. Gilgamesh is attempting to bridge immortality – there is no central divine figure or figures who are the fulcrum of his quest. He receives divine advice from Utnapishtim, but in the main his quest is a human quest. The gods are distant from Gilgamesh, almost watching in amusement, but certainly not central to his theology. Immortality is more important to Gilgamesh than theology. Qoheleth, on the other hand retains a faith in a central God – "God will bring every deed into judgment ... whether good or evil" (12:14; cf. 12:7, 12:1, 8:12).

Of course the main similarity that cannot be denied between the two texts is the heavy anthropological thrust. Humanity's inability to discern God's ways (Qoheleth) and find immortality (*Gilgamesh*) is central to both texts. But Qoheleth's skepticism runs deeper. Though Ecclesiastes is considered wisdom literature, Qoheleth questions the wisdom enterprise in its entirety, calling it an "unhappy business" (1:13). Qoheleth questions the ethical order of life, the linking of age and uprightness, the freedom of humanity, showing men and women their limitedness. [40] The admonition to enjoy life because of its brevity and chance, found in both passages, speaks to a kind of human fatalism that both authors certainly embraced. [41] *Carpe diem* is a frequent enough theme, but one that makes sense primarily in a humanistic context, not an overtly religious one. Certainly the three similar passages cited above are all centered around human life, and that is no accident, given that the two texts are so similar in their anthropological views. The limitedness of human life, the value of fellowship, and enjoying one's days are eminently humanistic themes. It is in their religious views where the main differences lie, and particularly in Qoheleth do we see these humanistic themes combined with a religious foundation. [42] Qoheleth's *carpe diem* can be seen as an embrace of creation, which is offered by God the creator to be "seized," or delighted in. Qoheleth must be given credit for bringing together *carpe diem* and theism in an astoundingly creative manner. Qoheleth unflinchingly affirms both the maddening solitude (*vis–à–vis* the divine) of human life, and the God who is in control of it all. This juxtaposition of themes accounts for Ecclesiastes' enduring esteem.

There are certainly other ancient Near Eastern passages that could be compared to Ecclesiastes and its "vanity of vanities" theme, [43] but let us move on to another comparison, that of Job and the Mesopotamian *I Will Praise the Lord of Wisdom*.

I Will Praise the Lord of Wisdom & Job

I Will Praise the Lord of Wisdom (*Ludlul bel nēmeqi* [44]) is an Akkadian work dated from between 1500 and 1200 bc. It has long been compared to Job, and the theme of a righteous sufferer was a prevalent one in the Ancient Near East. While there is no known direct link between this work and Job, because of our contention of common themes in the ancient Near East, these two works can still be exemplary for us to use in examining their shared perspectives. In the Akkadian story, the holder of a high social position loses his status and is afflicted with mysterious ailments. He cries out to the gods, but gets no answer from them. Even though he has done all that he is expected to do socially and religiously, he is stricken physically. The tale ends with a long recitation of how Marduk rescued him.

Indeed there are a number of similarities with Job. Here are a few examples:

> I, who used to walk like a proud man, have learned to slip by unnoticed.
> Though I was a respectable man, I have become like a slave…
> If I walk the street, fingers are pointed at me.
> (*Ludlul bel nēmeqi, ANE*, p. 150)
> He has put my family far from me, and my acquaintances are wholly estranged from me…
> Even young children despise me; when I rise, they talk against me
> (Job 19:13,18)

> [They all joined in] and came on me together.
> [They struck] my head, they enveloped my skull;
> [My] face is gloomy, my eyes flow.
> They have wrenched my neck muscles and made (my) neck limp.
> They struck [my chest], beat my breast.
> They affected my flesh and made me shake
> (*Ludlul bel nēmeqi, ANE*, p. 153)
> My adversary sharpens his eyes against me.
> They have gaped at me with their mouths;
> They have struck me insolently on the cheek;
> They mass themselves together against me.
> (Job 16:10)

Like the sufferer in *I Will Praise the Lord of Wisdom*, Job has his status taken away, is a mockery in public, and feels that he has done all that he can to be right with the divine. Job, too, feels that God has and is punishing him unjustly. Interestingly, a major similarity is that both hold a high view of the divine. *I Will Praise the Lord of Wisdom* opens with a praise to Marduk and ends with a praise of Marduk, the one who controls things and rescues the sufferer. Job as well, opens with Job, blameless (*tām*) and upright (*tummā*), fearing God and shunning evil, and in the end, repenting in dust and ashes before Yahweh. Job cannot fathom that the God who is just and righteous could be doing these things to him, though at no point does Job assume that God has abandoned him, as the writer of *I Will Praise the Lord of Wisdom* does with Marduk (*ANE*, pp. 148–149).

While there are certainly some *prima facie* similarities, there are also some glaring differences. The first, on a literary level, is that Job is a vastly more detailed, nuanced, pathos–driven account of an unjust sufferer. *I Will Praise the Lord of Wisdom* lacks the intimate divine perspective that brackets the beginning and end of Job. The real genre of *I Will Praise the Lord of Wisdom* seems closer to the lament psalms of David, where he complains about those chasing him, and he desires that God would hear his cause and restore him to his former place (e.g. Ps. 38). In fact, the situation of the sufferer in *I Will Praise the Lord of Wisdom* seems to be that he has been betrayed by another (*ANE*, p. 150), not that he is a righteous victim. In addition, *I Will Praise the Lord of Wisdom* does not explore the main theme of Job, which is "why" do the righteous suffer? This "why?" of suffering is the main axis of Job, while in *I Will Praise the Lord of Wisdom*, we see only the resolution of suffering in the final book. Moreover, the sufferer's relationship with Marduk is left unexplained. [45] The gripping agony of Job is missing in the Akkadian work. Finally, though there is much ascribing both the problem and resolution of the situation to the deity, the deities in *I Will Praise the Lord of Wisdom* are numerous, while the God of Job is singular and overall. Various gods, goddesses, exorcists, *et al.* are consulted by the sufferer, while Job's complaint is with Yahweh alone (and perhaps his "friends"). If there would be an overall summary comparing the two works, it would be that there are a number of superficial similarities, but that the depth of the work, and numerous important details, are different. Job, ironically, is the much more *human* work, exploring the crevasses and crags of the human psyche in suffering, while *I Will Praise the Lord of Wisdom* solves the situation rather one–dimensionally. Many have noted that its comparison with Job is inadequate, including W.G. Lambert who has suggested "The Babylonian Pilgrim's Progress" as a better title. [46]

Nevertheless, in *I Will Praise the Lord of Wisdom* we have a counterpart to biblical wisdom literature. Elsewhere in the ancient Near East, we encounter other theodicies. In Babylonia, akin to *I Will Praise the Lord of Wisdom* are two works, known as *The Babylonian Theodicy* and the *Diologue of Pessimism.* While *The Babylonian Theodicy* resolves the situation more or less orthodoxly, the *Dialogue of Pessimism* debates that orthodox view more robustly and reaches a conclusion that is openly agnostic and unorthodox. [47] *The Babylonian Theodicy* is a dialogue between the sufferer and a companion on whether piety ultimately is rewarded in this life. Ultimately, they both agree that the rich and powerful oppress the poor, and that the gods gave humanity a sinful bent at creation. The *Diolague of Pessimism* is a lively debate between a master and his servant, and is so absurd in content, that it has been interpreted as both comedy and satire (i.e. "Do not sacrifice, master, do not sacrifice. You get your god to follow you about like a dog" {*ANE*, p. 259}). The "vanity of vanities" theme is heavily prevalent in the *Dialogue* – in fact all courses of action are rejected except suicide in the final section. What we have with these two examples is an intensifying of earlier Babylonian work, and if all three were taken together, would be a worthy counterpart to Job. In Genesis 1, in comparison to the viewpoint of the *Babylonian Theodicy*, we also see humanity created as "very good." And in reference to the *Dialogue of Pessimism*, Hebrew writers never get to the point of saying all action is meaningless; in fact the farthest they are willing to go is that the point of life was to eat, drink, and enjoy one's life (Qoheleth), which in fact may be the antithesis of meaningless! We can agree with Lambert in that the prevalence of this theme suggests a common wisdom motif in the background of these works. [48]

What do we learn then from comparing Job with other ancient Near Eastern theodicies? First of all, the depth of human suffering is not explored to the depth that it is in Job, which may say something of how valued the human experience is in the Hebrew worldview. Secondly, none of the theologies match the singular monotheism of Job. The question may also be asked here, why didn't the Hebrew writers push to the edges of despair and nihilism that were around them in the literature? The answer lies in the base–level confidence (or fear) that they had in Yahweh. Though Job's situation may not have been resolved to explain all the issues with unjust suffering, in the end the Hebrew community was content to leave the result in Yahweh's hands. Yahweh's address to Job dispelled any doubts as to who was running the universe. Somehow, this theism, coupled with a robust humanism, kept the limits of orthodoxy for Israel.

The question arises here, with all of this discussion of literary dependence, of how the point can be made that Israel was borrowing from differing *religious* viewpoints. Is it not enough to say that Israel borrowed from the literary "soup" that surrounded her? I believe to leave it at that would be to miss a significant element in Israel's creating of her own sacred texts. These other texts, from Babylonia, Mesopotamia, etc., are part and parcel of these cultures' *religions*. We have the frequent references to Marduk in *I Will Praise the Lord of Wisdom*; the distant gods of *Gilgamesh*; the inscrutable god of *Amenemopet*. All of these are religious works in their own right – but of a different religion than Yahwehism. Yet the biblical writers included similar themes, even direct material from these other religious viewpoints. Now certainly the biblical writers were using common human themes present in all these different religions, but they were not shy about incorporating in their own theology different wisdom perspectives rooted in alternate religious viewpoints. This material was shaped, and re–emphasized, using Israel's own wisdom tradition as the template, and its own theology as its core. This, I believe, is a model for how to do contemporary Christian theology in a pluralistic context.

The Personification of Wisdom in Proverbs

"Lady wisdom is the most striking personification in the entire Bible", so says Roland Murphy. [49] This image is found in Proverbs 1,8 and 9, Job 28, Ecclesiasticus 24, Sirach 24, Baruch 3:9–4:4, and Wisdom 7–9, and so is a major theme for wisdom studies. There is much to mine for the theological picture of wisdom, but for our purposes we will concentrate only on the Proverbs passages. In particular, we are interested in the sources for Proverbs' usage of lady wisdom. Where did the imagery come from? Did Israel borrow from her neighbors, and if so, how? Wisdom gives an address in Prov. 1:22–33, and is contrasted with folly in Prov. 9, but Prov. 8 gives the fullest picture of lady wisdom. We will make a few introductory remarks about the content of this picture, then proceed to outlining some of the sources, which will in turn elucidate even more Israel's viewpoint of wisdom. [50]

Wisdom proclaims in public places to all people (wise and fools) that her way "is better than jewels" (Prov. 8:11), that kings and rulers reign by her

(8:15), and that riches and honor are with her (8:18). Her clarion call is then bolstered by reflection of her place with God at the beginning of creation (8:22). She was the first of his creation, and was present as he fashioned the universe – "when he established the heavens, I was there" (8:27). Not only was she present, she rejoiced in God's creation and was daily his delight (8:30). And finally, the one who finds her finds life itself, and receives favor from the Lord (8:35), while those who miss her find death (8:35). Here in these passages we see not only the remarkable elevation of a female personage (perhaps to or from divine status), but the clear importance of wisdom, and its crucial linkage to creation.

Where does this exceptional imagery come from? A widely used theory is to see personified wisdom as a Hebraized version of the Egyptian ma'at. Ma'at was sometimes personified as a goddess, and many parallels have been seen with this and Proverbs 1–9. [51] Von Rad claimed that, especially in 8:22–29, "the style of a specific Egyptian divine proclamation has clearly been borrowed, and that in vv. 30f. the Egyptian idea of a deity caressing personified truth (ma'at) has somehow, though not without internal modifications, found its way into our didactic poem." [52] Day has pointed out a number of problems with this view: 1) Ma'at is not personified anywhere in Egyptian wisdom literature; 2) When it is personified outside the wisdom literature, Ma'at never speaks in the first person or hardly at all, which contrasts strikingly with wisdom in Prov. 1–9. [53]

A second general theory of derivation is that of the goddess, either the Egyptian goddess Isis Sophia or the Semitic goddess Asherah. Concerning Isis Sophia, speculation about her dates from the Hellenistic period, and so is too late to have influenced Proverbs. [54] Regarding Asherah, Mark Smith notes that both Asherah and wisdom are feminine divine figures, that wisdom is associated with the tree of life (Prov. 3:18 – cf. the Asherah tree), and that the word 'ašrê, "blessed", occurs in the passages where wisdom is personified (Prov. 3:13, & 18). [55] However, Day points out that Asherah is never associated with wisdom, either in Ugaritic texts, the OT, or any other source. Furthermore the use of the word 'ašrê is unremarkable since it occurs quite frequently in Proverbs (Prov. 8:32, 34; 14:21; 16:20, etc.).[56]

There is also the possibility of a Western Semitic wisdom tradition influencing Proverbs. [57] The Words of Ahiqar read concerning wisdom, "To gods also she is dear. F[or all time] the kingdom is [hers]. In he[av]en is she established, for the lord of holy ones has exalted [her]" (ANE, vol. 1, p. 246).

Clearly wisdom is personified here. Regarding the pre–existence of wisdom, the famous passage in Ecclesiasticus 24 makes clear wisdom's preeminence – "I came forth from the mouth of the Most High, and covered the earth like a mist. I dwelt in the highest heavens, and my throne was in a pillar of cloud" (24:3–4). Thus another strong option is to see Israel as indebted to the West Semitic wisdom tradition here.

A final option is to see Lady Wisdom as a composite figure, perhaps culled from goddess literature or Western Semitic literature, perhaps with roots in the Egyptian ma'at worldview, but one that was adapted to Israel's viewpoint. As a transitionary figure on the way to completely identifying wisdom with the Torah (cf. Ecclus. 24:23), lady wisdom may be seen as a literary device. Judith M. Hadley sees "the apparent apotheosis of Lady Wisdom in the biblical literature [as] not a legitimization of the worship of 'established' goddesses, but rather [as] a literary compensation for the eradication of the worship of these goddesses." [58] Somehow the feminine needed to be expressed in the face of the gradual eradication (or assimilation into Yahweh) of legitimate goddesses such as Asherah. [59]

Irrespective of the motivation (the accentuation of the feminine, a teaching or literary device) or the origin (Egyptian, Semitic), we come ultimately to the theology of Lady Wisdom. How did Israel use this concept? What aspects of Lady Wisdom are similar or different from Israel's counterparts? The first and most obvious aspect to Lady Wisdom is her gender. That Israel assimilated a feminine figure into the central parts of her theology is remarkable given the patriarchal setting of the time. While the towering monotheism of Yahweh would not be called into question, still the presence of Lady Wisdom at creation points toward a singularly important role for the feminine in Israel's worldview. Secondly, we see the emphasis on "delight" – wisdom is God's "delight", and "delights" in the human race. [60] Wisdom was "rejoicing" (or "playing" – sachaq) before the Lord and in his inhabited world. This high view of play, or rejoicing, is significant, given the earlier Old Testament emphasis on Sabbath. [61] If we interpret the Sabbath as a feast for creation – as Moltmann does – a feast to enjoy the creator, to remember and preview the coming feast, then we see an important emphasis on playing or rejoicing. Thirdly, given Lady Wisdom's long comparison with Dame Folly (ch. 5–7), we see the moral/religious choice before Israel explained in sexual terms. These options are displayed on two levels [62]: 1) Personally – Wisdom is personified as a woman, and sexual seduction as Dame Folly; 2)

Covenantally – Israel has the choice of life or death before her, fidelity or infidelity to the Lord. The pungency of this analogy is significant given Israel's sad history concerning sexual behavior, and further serves to emphasize both the importance of sexual fidelity, as well as covenant fidelity. Fourthly, and perhaps most importantly, we see in Lady Wisdom a significant turn toward humanity. Von Rad sees Lady Wisdom as the personification of the world order inherent in wisdom literature. [63] He writes, "the most interesting feature of what is new is that this world order turns, as a person, toward men, wooing them and encouraging them in direct address." [64] The imagery of wisdom coming toward humanity, wooing men and women, is quite different than a distant God or gods unconcerned about humanity. How far apart is the static order of ma'at from the pursing figure of Lady Wisdom!

Finally, we may note that Lady Wisdom is a quite varied figure – "chameleon–like" as Murphy puts it. We have not analyzed Job 28 or Baruch 3:9–4:4, but we have enough of an idea of Lady Wisdom to see a multi–faceted portrayal. From being with God at creation, to delighting in humanity, to wooing humanity, to being identified with the Torah, Wisdom is a broad enough figure to outshine any ancient near eastern counterparts from where she may have originated. The breadth of her character warrants serious consideration as a major theme of Old Testament theology, and is a fitting place to end our survey.

Summary

It seems beyond dispute, despite the imprecision of tracing direct linkages, that Israel borrowed from her neighbors for her wisdom literature. There was a large deposit of literature available in that area and time. Only in a few cases (i.e. Amenemopet) is direct linkage or dependence clear. More attainable, however, is a picture of how Israel either crafted and shaped available material, or created new material. In terms of Israel's theological method we see a clear openness to ideas, and even theology, from other sources besides her own tradition. This is long before the advent of the ideas of general revelation or natural theology. In a sense, Israel was practicing pluralistic theology before the categories had even been established, so we cannot discuss her method anachronistically in terms of a natural theology. The method we observe in the most clear instance of borrowing (Proverbs and Amenemopet) is that Israel used certain material from other religious worldviews if 1) it corresponded with (or perhaps did not contradict) material already present in the Israelite tradition; or 2) it provided a way to emphasize distinctive elements of the Israelite tradition. Further in the character of Lady Wisdom, we see a new wrinkle illumined for Old Testament theology – the feminine figure of Lady Wisdom. Ultimately, however, the theological method boils down to the theology employed – Israel desired to emphasize certain theological points in using foreign religious sources. So in this concluding section, we will summarize some of the theological distinctives that guided Israel in writing her wisdom literature in theology proper, and anthropology.

Theological Distinctives

Everywhere we turn in Israel's wisdom literature we encounter Yahweh in control. In Job, God is the one behind the divine "test", and as the one who answers convincingly from the whirlwind. In Ecclesiastes, though Qoheleth doubts that God is knowable, ultimately he retains belief in the God who "will bring every deed into judgment" (Eccl. 12:14). In Proverbs, God is the sure and steady force behind the judgment of good and evil. This view is similar to that found in the Ancient Near East: In Egypt ma'at is the ever–present force; the gods of Gilgamesh are similarly in control of all the reality around the hero; in I Will Praise the Lord of Wisdom Marduk is seen as the overseer of creation.

But how Yahweh is in control is where Israel parts ways with her neighbors. Yahweh is the motivator for good deeds in Proverbs, not the distant God who judges human activity. Yahweh imbues ethical and moral behavior with a religious quality missing in Egypt and Mesopotamia. There the gods wait to judge and then render judgment, acting out a deed/consequence construct. Yahweh also works as an advocate – for the poor. This sense of personalness is even seen in Ecclesiastes, where Yahweh cares about every deed of his adherents, as opposed to the lack of central divine figures in Gilgamesh. Yahweh as involved and personal is taken perhaps to its highest level in the personage of Lady Wisdom. Here we see uniquely the divine coming toward humanity, wooing creation, enjoying and delighting in it. Von Rad suggests, "Still, the most important thing is that wisdom does not turn towards man in the shape of an "It," teaching, guidance, salvation or the like, but of a person, a summoning "I." So wisdom is truly the form in which Jahweh makes himself present and in which he wishes to be sought by man." [65] Lady Wisdom also highlights the feminine role of the divine. Not content with a solely patriarchal view of God, Israel augmented her view with female wisdom. Further, the divine figure of Lady Wisdom rejoices and plays, [66] a powerful image for Israel, one that provides a strong counterpoint to a somewhat austere monotheism. Here we may see an aspect of mediation that finds its fruition theologically in the Christological mediation of the New Testament. Lastly, Lady Wisdom provides strong links to the Sabbath and creation.

In short, the theological picture that emerges from Israel's contact with other nations and religious systems in her wisdom literature is one that highlights the very essence and breadth of her revelation: A monotheism that is personal, a powerful God that works for the weak, a mediatory figure that enjoys humanity. It is in these three areas that Israel sought to establish herself theologically with respect to her neighbors. All of these serve to underscore those elements of Israel's religion that were crucial for her own self–understanding, her distinctiveness, and her survival. Israel was not concerned about where the material needed to frame her worldview came from – it did not have to be Mt. Sinai – only that the material, and her usage of that material, highlight areas of thought that were either neglected, needed for a particular time, undeveloped, or all three.

However it is still important to note that these three areas were present in some form or fashion in the Torah. Yahweh called Moses personally, rescued the relatively insignificant Israelites from powerful Egypt, chose Abraham as the person through whom the world would be blessed, and created the Sabbath as the day to enjoy the creation he pronounced "very good". Although these themes were present in Israel's tradition, somehow they needed to be re–emphasized because of neglect or the exigencies of Israel's circumstances. Israel felt secure enough in her tradition to borrow from other traditions, and this is key in developing our own methodology for today. In that borrowing we see an enrichment of Israel's tradition – for example, the figure of Lady Wisdom is striking and profound, and may not have been in Israel's tradition at all save for the presence of neighboring religions.

As strong as Israel's theological distinctives were, Israel's humanity is also lifted up through the questioning of Qoheleth and the carpe diem theme. So now we must turn specifically to that anthropology.

Anthropological Distinctives

The other basic category on which Israel put her own stamp is anthropology. Certainly in a few pointed instances, Israel emphasized the human aspect of life in a previously unexplored manner. The complaints of Job and the resigned humanism of Qoheleth both exemplify the growth of an anthropological perspective in religious literature that is unique. But beyond that, some specific motifs were appropriated from other wisdom sources while others were rejected as Israel developed her own viewpoint. In the case of Amenemopet, Israel seems to have mirrored the heated/silent man with the wise/foolish man, even while giving the latter more of a religious hue. The wise or foolish man is directed to "fear the Lord," whereas fate and fortune are key for the heated or silent man. The wise/foolish man's behavior is also linked with an involved God as opposed to distant gods or an impersonal ma'at. In both cases, the anthropological portrayal serves to stress an aspect of Israel's theological distinctiveness. At the same time it also reinforces creational themes consistent with Israelite theology.

A theme that is rejected in Old Testament wisdom literature is that of an impersonal fate and fortune as being central to ethics. Perhaps it is flirted with in an oblique way in Ecclesiastes, where carpe diem is a main subtheme. Israel joins her neighbors in affirming the brevity of life, the value of fellowship, and the enjoyment of one's days. But even there, Qoheleth is concerned to maintain a God that is involved both in judging everything and giving gifts to humans. Atheism, or even a kind of agnosticism, is not seriously entertained in Ecclesiastes. Certainly, humanity has a difficult time discerning the will of God (Israel concurs here with her neighbors particularly in the sometimes inaccessibility of the divine [cf. Job]). But there remains a personal, religious quality to the ethics of Qoheleth and Job that is lacking in Gilgamesh and I Will Praise the Lord of Wisdom. Gilgamesh is more concerned with immortality, while Job's concern is with righteous behavior before Yahweh. This may again be the result of human reflection on distant polytheism as compared to a vibrant monotheism.

Looking at the theodicies of the Ancient Near Eastern wisdom literature shows us that Israel explored the pathos of human existence in better detail than her counterparts. Israel seems to take suffering more seriously and asks harder questions of God. But she does not succumb to the nihilism or fatalism that permeated her neighbors. This may be explained in part by Israel's belief that humanity was created good, not given a sinful nature at creation. It also might be explained by the singular monotheism that was maintained in the face of polytheism. This heightened humanism, alongside a reinforced monotheism, enabled Israel to explore the edges of humanity further, even encouraging an enjoyment of life, rather than progressing to the fatalism or supernatural "fixing" that seemed to result from the distance inherent in neighboring religious systems.

Finally we may note that Israel's covenant fidelity is not utterly lacking in its wisdom literature, so that ascribing a purely anthropological perspective to wisdom is too limited a perspective. In the figure of Lady Wisdom we see both a model for morality and a device for putting morality in the ultimate perspective: life or death; adherence to God's ways or not. Lady Wisdom exemplifies the personalness of Yahweh, the quest of humanity to find her being outlined in such personal terms. Humanity is not to do right simply because God requires it. They are encouraged to fall in love with wisdom, to pursue her and gain her as a gift from God. [67] Thus Israel built a heightened humanism and a vibrant monotheism in part through the religious works of her neighbors, even as it transcended or modified these other religious viewpoints in important ways.

Conclusion

Until more literature is discovered, hard and fast linkages between ancient Near Eastern wisdom literature are challenging to prove. Nevertheless, we know enough about the time and genre to note a huge body of literature, a general "soup" that was present in the entire area, available to be adapted to each culture, and some specific "borrowings" that seem likely. Israel certainly partook of the wisdom genre, but in her own unique way that emphasized her own unique religious worldview. As we have already noted, Israel used material from other religious worldviews if, first of all, it corresponded in some way with material already present in the Israelite tradition. This material did not have to agree exactly with Israel's tradition (cf. ma´at notions), but it did have to have a certain affinity with Israel's tradition so that it could be shaped to fit Israel's needs. In the process, Israel found herself not only taking material from other traditions and adapting it to her own religious scheme, she was learning from others (cf. Lady Wisdom). Israel did not attempt to integrate all of the available material from other religions into her theology. How did she choose which ones to use? This unfortunately may be a question that is not answerable from where we sit today.

The second part of our analysis of Israel's wisdom hermeneutic is that material was included from other religious viewpoints if it provided a way to emphasize distinctive elements of the Israelite tradition. In our particular study we see the three themes of a personalized monotheism, a God working for the weak, and a mediatory figure enjoying humanity emphasized on the theological end. Anthropologically, we see the human quest for wise behavior given a strong religious and even personal framework, and the pathos of human life given more depth. The themes that we might encounter in other religious viewpoints today would be different, given our vastly different eras. But the integrative process would be the same, if we follow the example of the Israelites who chose pluralistic integration under the construct of their own tradition.

The third option to Israel's borrowing has to do with the illumination of new areas of theological expression. We see this primarily in the figure of Lady Wisdom. Not content with a singular, patriarchal monotheism, Israel strove to integrate a feminine, mediatory figure into her theology. This arena of innovation, of adaptation was crucial for Israel to fill out her own revelation. In our era, the theological equivalent would be integrating new areas of general revelation with our special revelation core. This area may illumine previously neglected areas of theology, or provide innovation under the aegis of the Spirit.

The core of Israel's tradition was creation and the creator. God was over all creation, and thus creation provided the framework for assessing outside revelation and competing ideas. At every turn in wisdom literature we see creation popping up as an important theme. In Proverbs, Lady Wisdom, who was with God at creation, calls humanity to follow her. In Ecclesiastes, both "vanity of vanities" and the refrain to "enjoy life" are rooted in creation and creator, and the final chapter begins with the admonishment to "remember your creator in the days of your youth". In Job, Yahweh, answers Job out of the whirlwind, asking, demanding if Job was there when he laid the foundation of the earth. His evidence and questioning is rooted in nature and its marvels. It seems that at the significant junctures of Israel's interaction with her surrounding neighbors she found in her doctrine of creation a means of engagement. Moreover creation provided the yardstick by which Israel could measure these ideas and innovations. Israel's theology of revelation was broad enough to include ideas from other traditions, but always the backdrop of Israel's own tradition guided it. This tradition, though, wasn't made up only of holy texts, but rather of the community thinking theologically about life and God, engaging Israel's radically different neighbors in issues of morality, ethics, and worship.

In our day, Christians have the scriptures and thousands of years of history to guide them. But we live in a pluralistic, postmodern world. Possessing that tradition must not limit us from thinking theologically about life and God, including examining what we can gain from other religions that might enhance or deepen our own revelation. This material need not agree completely with our own tradition (cf. ma'at). However we must continually measure these contributions against the contours of our own tradition. In wisdom literature we have an example of that process that can serve as a tutor as we attempt to do in our day what Israel did several thousand years ago.

A Model

As we close this chapter, let us summarize where we have come thus far in developing our model for how to do a postmodern, pluralistic Christian theology. We have outlined our pluralistic context, set forth key epistemological and theological rationales in postmodernity and in natural theology, respectively. Central in natural theology is the role of the Holy Spirit as the thread that connects natural theology with "revealed" theology, general revelation with special revelation. We then explored two major theologians in the area of creation, one open to other religious thought – Ninian Smart, and one staying within the confines of traditional Christian doctrines – Colin Gunton. There we found what has seemingly been common in these discussions, and that is the taking of either of one or two roads: The first is that by being open to other religious traditions, one's theology tends to be so pluralistic that it does not reflect a (Christian) tradition well; the second is that the avoidance of any other religious thought results in a closed viewpoint that has difficulties in participating in the global, diverse, "tournament of narratives". Our task has been to provide a third way – one that is consciously Christian and still a partaker of the global theological community.

If we are going to pursue theological resources outside of the Christian tradition, are there any guidelines for doing so to be found in the history of Christian thought? The answer is, not many. Not surprisingly, only recently do we find scholars attempting to integrate material from other religions into a Christian viewpoint. In the end, this is puzzling, given that we find right in the Bible a model for how to use other religion's doctrines. Old Testament wisdom literature shows us that an idea from other religions may be used in a Christian doctrine if 1) it corresponds with material already present in the Christian tradition; 2) if it can be used to emphasize distinctive elements present, but perhaps neglected in the Christian tradition; and 3) it can illumine new areas of theological expression. If a pneumatological core is guiding our sense of revelation, we can move beyond reinforcement and correction to learning something fresh, i.e. something that not only doesn't contradict the Christian tradition, but extends it. Can we push not only the "use" aspect, but also the "learn from" aspect of pluralistic theology? I believe we must. Our pluralistic context is pushing this generation to consider how God's revelation is operative in a global, interconnected world. And there are aspects to that general revelation that can enhance and extend the depth of our own theology.

In any case, we have come now to the place where we can put this journey we have been on to use. Now that we have established that we need to do what we're doing, and how we can do it, and why it is necessary, we must proceed to do pluralistic Christian theology. Here is then an outline for our model that we will use in exploring a pluralistic Christian doctrine of creation:

- **Discovery** – find source material in original sources of other religions.
- **Similarity/dissimilarities** – locate similar themes and/or challenges to the Christian story.
- **Theological guideposts** – what are the theological guideposts within Christianity that we can refer to in discussing these areas?
- **Adaptation** – adapt the material into a Christian theology, noting the level of adaptation, which are listed here from strongest to weakest:
  1) adaptation – bringing in new ideas more or less as a whole.
  2) assimilation – modifying the idea for Christian theology.
  3) absorption – enfolding an idea into an already existing Christian doctrine.
- **Presentation** back to the larger world theological community

This is the model we will follow in the next few chapters, first, as we explore cosmogonies in other religions, then as we return to craft seminal areas of a pluralistic Christian doctrine of Creation

CHAPTER SIX

Cosmogonies in Other Religions & the Christian Doctrine of Creation

Thus far we have described the pluralistic context for our enquiry, established an epistemological and theological rationale, noted both concerns and possible directions for a Christian doctrine of creation through our two representative samplers, examined historical theology for examples, and developed our own model from Old Testament wisdom literature. As we continue on our journey, [1] it remains for us now to do what we have been advocating: examine source material from the three other major world religions – Buddhism, Hinduism, and Islam – not only as dialogue partners with a Christian doctrine of creation, but also to see if there are any areas we might incorporate into our own doctrine. The key question for us here is what are the salient points regarding cosmogony in Buddhism, Hinduism, and Islam? Now certainly it must be recognized at the outset that to summarize, must less creatively engage, an entire religion's perspective on a matter in such a short space, is a daunting, if not impossible task. It is also clear that an analysis of a religion's doctrine must not be limited strictly to the authoritative texts, for that methodology may be an imposition of a Western tradition on non–Western viewpoints. [2] The ideal would be an *emersion* into a tradition to soak up its outlook, what W.C. Smith advocates as a relational as opposed to a propositional view of truth. [3] But nevertheless, the task before us in this book is primarily a textual one, and one that could be exemplary in that any examination of a religion's (other than Christianity) texts and tradition is to be preferred over generalizations and outright ignorance when it comes to analyzing the Christian story in its world context.

We will center on issues of cosmogony and cosmology, and the implications thereof – mainly in the areas of the nature of God, humankind, ontology, and preservation of the universe. In terms of cosmogony, there are several options available – creation (here we are aware that the use of the word "creation" assumes many facets that some systems may not include as a part of their worldview), struggle and victory over opposition, struggle over chaos (i.e., Tiamat), cosmic egg and cosmic world parents, etc. [4] Rather than outlining these options generally, we will let each tradition speak for itself. In terms of cosmology, we are interested not only in the set–up of the world and universe, but the implications therein for a worldview; i.e. what does the physical genesis and placement of the universe tell us about a tradition's viewpoint of humanity, of God, etc. [5]

As we look into Buddhism, Hinduism, and Islam, we will include a brief description of the religion before moving onto key points that relate to a Christian doctrine of creation. We will conclude with a summary of points of connection and disconnection with a Christian doctrine of creation, as well as a deeper look into three areas that hold promise for pluralistic theology. Our ultimate goal is to see what, if any, areas of correspondence can be made between Christian theology and the three other faiths, and so provide springboards for dialogue and incorporation in our subsequent discussion. Our pluralistic, postmodern era requires both this exploration of other faiths as well as the clarification of our own tradition that results from this inquiry, if our theology is to be culturally relevant.

Buddhism

Maseo Abe sets down for us a good comparison between Buddhism, Islam and Christianity: "At the heart of Christianity is a person – Jesus Christ. At the heart of Islam is a text – the Qur'ān; at the heart of Buddhism is a story – that of Buddha's realization." [6] And so any attempt at synthesizing Buddhist thought must begin with the quest–like nature of Buddhism, for it is not so much concerned with orthodoxy as orthopraxy. In fact, Buddhism turns many of the core Western religious beliefs on their heads – for example, there is no God to believe in, and the existence of a person or a thing is ultimately denied. Buddhism itself is to be discarded in pursuit of the ultimate, which is nirvana. This pursuit begins with the four great truths taught by Buddha, who was not only a historical figure but also a type: 1) the existence of suffering; 2) the cause of that suffering is desire; 3) that cause can be removed; and 4) a blueprint of how to remove that cause is available. This blueprint is known as the Eightfold Path, which includes "right view; right aspiration; right speech; right action; right livelihood; right effort; right mindfulness; and right meditation." [7] This eightfold path leads to nirvana, but these truths are not the essence of Buddhism, for they, too, are to be discarded once one has reached the ultimate.

This ultimate reality is not God, or being, but *Śūnyatā*, or "emptiness", which denotes the absence of any kind of self or selfhood. [8] This ontology of emptiness points to a key concept in Buddhism, that of inter–relatedness. Things exist not on their own, but only in relational contingency to others. Moreover, there are two realities, a dualism between what we see, conventional truth, and what really is, ultimate truth. This has particular implications for anthropology. The human soul in Buddhism does not exist permanently, in contrast to Hinduism's reincarnations of the soul and Christianity's view of the soul.

The Buddhist profession of faith involves, then, three "treasures": 1) Go for refuge to the Buddha; 2) Go for refuge to the doctrine (*dharma*); 3) Go for refuge to the community. [9] Soteriologically speaking, salvation is to be found in the here and now, in teaching rather than revelation, in one's own experience and not in God, and in moral excellence ("skillfull means"). These "skillfull means" keep the emphasis in Buddhism not on faith, but on knowing and doing, though they are never to be construed as "the way" to nirvana, but only as a provisional path by which one achieves emptiness.

More of the specifics of Buddhism will be brought out as we examine the particular issues of cosmology and cosmogony, so to those areas we turn.

Key Issues in Buddhism Relating to a Christian Doctrine of Creation

The literature (sacred texts) in Buddhism is immense, which speaks to Buddhism's inclusive and expansive nature. But we must begin by describing Buddhism's cosmology and cosmogony. According to the seminal work in Buddhist cosmology, W. Randolph Kloetzli's *Buddhist Cosmology*, there are three broad categories of cosmologies: 1) the single world system or "*cakravāla*–cosmology"; 2) the numerological or mathematical cosmologies focusing on the thousands of worlds and innumerables of "Buddha–fields", and 3) the cosmologies of the Pure Land sects which concentrate on the drama associated with a particular Buddha–field, generally *Sukhāvatī*.[10] Most scholarship has focused on the single–world system, as descriptions of it are more readily available and coherent; the other two cosmologies have been all but neglected. [11] We will focus our comments on the first two cosmologies. [12]

The single–world system is perhaps the most prominent Buddhist cosmology, and is significant in older Buddhist texts. It can best be understood through the most systematic treatise on the issue, Vasubandhu's *Abhidharmakośa*, from the 4th or 5th century AD. The most basic way of describing the cosmology would be the cosmos as a flat disk, with heavens and meditation realms above and hells below:

> Here is how it is thought that the receptable world is arranged: at the bottom there is a circle of wind, immeasurable, with a height of sixteen hundred thousand leagues... By the predominate actions of beings, there falls from massed clouds, upon the circle of wind, a rain the drops of which are like the shafts of a carriage. This water forms a circle of water, with a thickness of eleven hundred twenty thousand *yojanas*... Then there is above the circle of water now reduced to eight hundred thousand *yojanas*, a sphere of gold, three hundred twenty thousand *yojanas* thick... Nine great mountains rest on the sphere of gold. In the center there is Meru; concentrically, the other seven are arranged around Meru; Nimindhara forms the exterior rim that envelops Meru and the six inner wall–mountains – whence its name. Beyond [Nimindhara] lie the four continents. Enveloping all is Cakravāḍa, thus named because it encircles the universe with its four contenet and also because it has the form of a wheel... Seven mountains are made of gold; the last is made of iron; and Meru is made of four jewels. [13]

This golden earth, or disk, is ringed with seven circular, golden mountain ranges, with Mount Meru at the center and the *cakravāla* wall of iron surrounding it (the *cakravāla* is the designation of this cosmology in Buddhist texts). On this land mass are four islands representing the four points of the compass. These islands have differing sizes and shapes regarding the lifespan of their inhabitants, the appearance of a Buddha, and the hells beneath.

Our interest here, though, is not in the specific geographic details of the cosmology, but in its implications. The universe is limited and specific, and the soteriological drama consists in the adherent traveling throughout the universe in the course of his meditations, eventually getting beyond it, and detaching himself from it and so entering nirvana. The presence of a Buddha is not central to this drama, as the central issue is the progress of the *arhat*, or Buddhist saint, away from the meditation realms toward nirvana, and the eventual extinction of nirvana itself. Here we have the beginnings of some central differences between a Christian cosmology and a Buddhist one in the issue of divine agency.

The second broad cosmology is that of the "cosmology of thousands", so named because of the thousands, even millions of worlds involved in this cosmology. There are countless variations of this, but a key aspect to this cosmology is its emphasis on infinity: The *Mahāvastu* considers that the number of chilicosms, or 'fields of Buddha,' is infinite in every direction:

> Pay heed to these words of mine and to my teaching. "A Buddhas' field is proved to be sixty–one systems of three thousand worlds …
>
> In the eastern quarter of the world there is a well–laid–out Buddha field, where abides the supreme Conqueror named Mrigapatiskandha.
>
> In the eastern quarter of the world there is a healthy Buddha–field, where abides the Conqueror with the thirty–two marks, named Siṃhahanu.
>
> In the eastern quarter of the world there is a resplendent Buddha–field, where abides the all–seeing great Seer, named Lokaguru.
>
> In the eastern quarter of the world there is a secure Buddha–field, where the Master named Jñānadhvaja teaches men.
>
> In the eastern quarter of the world there is a bright Buddha–field, where abides the Conqueror, named Sundara, who is radiant like the gold bimba.
>
> In the southern quarter of the world there is a Buddha–field that is full of palm–trees. There abides the Buddha named Anihata, who is the joy of devas.
>
> In the southern quarter of the world there is a pleasant Buddha–field, where abides the great Seer, the Buddha Cārunetra.
>
> In the southern quarter of the world there is a Buddha–field free from all impurity. There abides the Guide, the Buddha named Mālādhārin.
>
> In the western quarter of the world there is a Buddha–field that is free from strife. There abides the Buddha, the destroyer of existence, named Ambara.
>
> In the northern quarter of the world there is a pleasant Buddha–field, where abides the Buddha named Pūrṇacandra, learned in the sacred lore.
>
> In the nadir of the world there is a securely–fixed Buddha–field, where abides the Buddha, the Tathāgata, named Dṛiḍhabāhu.
>
> In the zenth of the world there is an unshaken Buddha–field, where abides the Buddha named Mahābhāga, a destroyer of his foes.
>
> There are besides thousands of other Buddha–fields, and yet other thousands, of which one cannot reach the end in enumerating.
>
> Thousands of empty Buddha–fields which one does not know where to begin counting, and thousands of universes of three–thousand worlds. [14]

The emphasis on spatial infinity in this cosmology does not preclude the importance of time and the appearance of a Buddha within this system. However, more central in this cosmology is the soteriological drama – how the adherent reaches nirvana. Kloetzli has outlined the main soteriological emphasis of this cosmology:

> While the soteriological drama associated with this cosmology is framed by the speculations on cosmic time, the drama proper divides itself into four discrete "moments." The first is that of the progress of the *śrāvaka*, or one who has undertaken the religious vocation toward becoming an *arhat*. Second is the exercise of miraculous powers. Third is the career of the *bodhisattya*, who makes a vow in the presence of a Buddha to pursue Buddhahood rather than pass into the extinction of nirvana. The fourth moment in the drama is the appearance of a Buddha. [15]

The interesting point for our purposes is that while Buddhism has an elaborate cosmology, it has little or no cosmogony, no doctrine of creation, *per se*. What it does have is a concept of "dependent co–origination", whereby everything arises together and in relation to another. Here we have not a rival conception of cosmogony to the Christian story, but a "non–creation". This concept links the origination of things with their end and with the ultimate reality in Buddhism – emptiness. Nāgārjuna states, "Never are any existing things found to originate from themselves, from something else, from both, or from no cause." [16] In fact, speculation as to the beginning of the world was discouraged – "The Buddha refused to answer these questions because they were vain and had no significance for salvation." [17] De La Vallēe Poussin writes,

> It is probable that a large number of Buddhists, imbued with the 'moralism' of their master, avoided frivolous curiosities – 'non–Buddhist' 'mundane' disciplines (*lokāyata*), – and were content with denying, on the one hand, a supreme personal creating power (a lord, *Īśvara*), against the theists, the Brāhmans; and, on the other, the innate independent power of things (*svabhāva*), against the *svabhāvavādins*, the materialists. The formula, 'The diversity of the world comes from the act,' contains for a well–informed Buddhist the Alpha and Omega of the necessary cosmological information.[18]

However, this supposed "non–interest" in cosmology or cosmogony must be challenged, if only for the fact that the sheer number of cosmologies indicate something significant for the Buddhist tradition. Winston L. King notes,

> Theravada Buddhism is almost unthinkable without its cosmology ... Should one attempt to remove it by some sort of theological surgery, the remaining corpus might well be a corpse instead! And so strong is the rootage of this cosmology in the tradition that the majority of orthodox monks and laity even today would resist any attacks upon or alterations in the Mount Meru cosmology (i.e. the "cakravāla–cosmology"). For them this cosmology represents the clear teaching of the Buddha as interpreted by the orthodox commentators, and this is the end of the matter. [19]

So while excess speculation as to cosmology and cosmogony may have been discouraged, perhaps because of the lack of a central divine figure, still the important role of the "beyond" in Buddhist thinking remains.

Buddhism replaces a God who holds things together with the interconnectedness of the universe itself. A Buddhist ontology would certainly include this concept of interconnection, analogous to the Christian doctrine of providence. Leo D. Lefebure brings out these differences between the Buddhist and Christian systems well:

> The difference between the Christian tradition, which looks to God as Creator, and the Buddhist nontheistic vision of dependent co–arising is especially important here. For Buddhists, dependent co–arising is the web of mutual interrelationships that binds all beings into a whole and makes them one. The Buddha's intuition of dependent co–arising is an intuition of pure contingency and of universal connection. All things are what they are, arising from the other things of the universe. Nothing arises from nothing, and so there is no question of a Creator calling things to be out of nothing. [20]

Abe points out how this interconnectedness leads again back to emptiness: "But if everything in the world depends on something else for being what it is, then nothing in the universe can be said to possess *svabhāva* or genuine self–existence; hence it is empty." [21] Co–origination is crucial for the Buddhist tradition, and might be thought of as the "cosmogony" within the system.

An implication of co–origination is perhaps the most provocative aspect of Buddhism as it relates to cosmogony and creation, and that is the lack of a divine agent altogether. Though Buddha is imbued with heroic, or exemplary stature, in no way is he seen as a creator, a sustainer, or even the most crucial aspect of the religion; his appearance is, to borrow a phrase which occurs throughout Buddhist literature, as rare as the blooming of an *udumbara* plant.[22] The ultimate reality is nirvana, not God. This is a radical counterpoint and challenge to the Christian tradition, for which a fundamental assumption is belief in God, the creator of all things. [23] It should not be assumed, however, that this lack of divine agent excludes transcendence from the Buddhist system. Indeed, Khin Maung Din writes, "The oriental refusal to predicate the Transcendence with western philosophical categories should not be interpreted as the denial of the Transcendence itself. This applies to Buddhism also. Therefore the Buddhist denial of the 'existence' of 'God' as a 'Personal Being' or 'a Creator' is not a total rejection of the indescribable, transcendental Theos." [24] Din puts forth an alternate construct for understanding God via a Buddhist mindset. Taking into consideration the Buddhist impulse to not describe ultimate reality at all ("the predication of Reality in human terms will distort and relativize its true nature"[25]), Din suggests we can learn from 'Oriental methodology': "Should a Christian theology continue to keep on referring to God as 'a Person,' 'a Father,' 'a Him,' 'a Creator,' etc., in an absolute sense? Should we not rather say that somehow God is a Person, as well as not–a–Person, that God is a Father as well as not a Father (a Mother?), that God is a Creator as well as not a Creator, that God is a Thou as well as not a Thou?" [26] This intriguing reminder of the non–absolute nature of theological language points us to some connections between Buddhist and Christian viewpoints in a doctrine of creation. What are the implications of a God who is "wholly other"?[27] Can we grasp the divine essence? And while in the Christian tradition we must assent firmly to existence (over non–existence), the tenuous nature of reality, of which the Buddhist tradition is very cognizant, is something that must be addressed in Christian theology. In fact, this connection between Buddhism and Christianity regarding the doctrine of God has been explored through the Christian concept of *kenosis* in detail by Maseo Abe, John Cobb, and others in *The Emptying God* (1990). [28]

Buddhism's two–tiered ontology of reality and non–reality is a common theme in Eastern religions, and has particular manifestations that touch on creation. It has already been mentioned that reality, existence itself, is to be left behind in the quest for nirvana, which is nothingness. For the Christian tradition, however, a full–orbed doctrine of creation must include the New Testament teaching of a "new creation", which leads toward the area of eschatology. Jürgen Moltmann, in particular, has emphasized the eschatological contours of a doctrine of creation: "If [the world] is understood as God's creation, the universe is caught up into the history of God's rule. Creation in the beginning prepares for this history, and the history itself is consummated in the new creation in the kingdom of God." [29] A Trinitarian conception of creation will include not only the Christological and the Pneumatological, but also the eschatological dimensions of creation. And this Christian picture of the end – God reigning over his kingdom – is quite different than the Buddhist one, absolute emptiness.

The Buddhist conception of reality (and non–reality) also touches on a crucial aspect of anthropology: The self in Buddhism has no significant role. Walpola Sri Rahula, a well–known adherent to the Theravāda Buddhist tradition writes, "Buddhism stands unique in the history of human thought in denying the existence of such a Soul, Self, or Ātman. According to the teaching of the Buddha, the idea of self is an imaginary, false belief which has no corresponding reality, and it produces harmful thoughts of 'me' and 'mine,' selfish desire, craving, attachment, hatred, ill–will, conceit, pride, egoism, and other defilements, impurities and problems."[30] As such, the self does not exist in Buddhism to be repressed (as in some Christian conceptions), rather it simply does not exist. In contrast to Hinduism, where the soul passes from one body to another, Buddhism asserts that there is no permanent element such as a soul in a person. What we do encounter in Buddhism is a strong emphasis on knowing and doing – the practical. Faith is not essential for Buddhism, therefore neither is a soul. Moreover, right faith, or orthodoxy, is not crucial either, but rather moral excellence, or orthopraxy, is valued. The central element for Buddhist anthropology then, springing as a consequence from its ontology, is the moral, or ethical.[31]

Contrast this "nonexistent self" with the Christian conception of the *imago dei*, a key part of a Christian doctrine of creation. The Christian doctrine, while avoiding Platonic extremes regarding the existence and overvaluation of the soul, asserts that humans are made in the divine image, and as such are valuable as individuals and as a part of creation. Without delving too deeply into exactly what the *imago dei* constitutes, we may safely say that while it includes a moral/ethical component, its pivotal core is correspondence, either in relationality or nature, to the divine.[32] Theology and anthropology are linked, and as a consequence, the Christian individual is imbued with existence and significance, both individually and corporately. That is the major point of difference between the two systems, qualifying any similarity based on a common concern for practical, moral living.

Finally, it would seem that with the lack of a divine subject, a nonexistent self, and the interconnectedness of the entire universe, Buddhism would fall squarely in an immanent, if not monist, category concerning divine ontology. As such, we would have major differences with the Christian tradition. However, Abe argues that Buddhism transcends all duality, including the transcendence/immanence dimension. Whereas in Christianity the ultimate is God, defined as supreme good, in Buddhism, the ultimate is neither good nor evil, but rather emptiness.[33] In our view, however, Buddhism cannot escape the problems associated with relating the ultimate or infinite and the particular or finite simply by dissolving and discarding the two on the way to nirvana. Paul Knitter posits that it does: "While *Sunyata* truly *is* every finite reality, which each particular entity truly is (and is not just 'a piece of') *Sunyata*, still 'particularity really exists.'" [34] We remain skeptical. While the ultimate (nirvana) is beyond everything, it still remains at its core "nothingness," which the adherent must pursue alone, and which by its very definition is not connected with finitude. However, this sense of disconnect is so sharp, that one wonders if there is any relation at all between the ultimate and the finite in Buddhism. There is no immanence in the ultimate that can connect it with the finite. In the end, the ultimate ends up being radically transcendent. And while the finite world is connected within itself, it remains radically disconnected to transcendence. So that in Buddhism, we encounter a system that is fundamentally immanent while at the same time on another level fundamentally transcendent. There seems to be no connection between the two, no mediation, which is such a crucial concept in Christianity.

The Christian tradition insists not only on transcendence – God, the ineffable Subject – but on immanence as well – through the persons of Jesus Christ and the Holy Spirit. But an emphasis on one does not exclude the other; in part, a robust Trinitarianism joins them together. This ontology allows for the interplay between distant and near, connection and disconnection, subject and object. Trinitarian thought helps bridge the gap between unity and difference, the universal and the particular. We have already noted Gunton's argument that the doctrine of the Trinity, with its epistemology of the one and three, provides a needed answer to the problems associated with transcendent and immanent categories. Other pluralistic theologians have also noted the crucial role the Trinity plays not only in Christian epistemology, but also in providing a bridge for dialogue with other religions. [35] Christian theology thus attempts to relate, through the mediatory roles of Christ and the Holy Spirit, the transcendent and immanent.

Now certainly, the Buddhist perspective forces the Christian to look further into the immanent side of the continuum as it relates to God. This may be a needed corrective to a too–transcendent portrayal of the Christian God, as Moltmann and others have argued. Knitter suggests, "To experience, with the Buddhist, the immanence of the Absolute is to discover – or rediscover – that the relation between God and world–humanity is not monistic (simple identity) or dualistic (clear distinction) but *non–dualistic*.... The Cartesian 'cogito,' with its radical subject–object distinction may be a profitable starting point for the natural sciences, but it constitutes a 'false start' for our understanding of the Ultimate." [36] A re–emphasis on the immanence of God may help us to not overemphasize humanity's ability to understand the ultimate, which Christian theologians may have been ironically guilty of in emphasizing the transcendence of God. By starting with the human/divine discontinuity, the human is overvalued, and his or her connections with the rest of creation are undervalued. In fact, Christian theology can agree with Buddhism in rejecting an anthropocentric view of the universe, while still advocating for the implications of a robust sense of an immanent understanding of God.[37]

While the immanence of God and the littleness of humanity are important themes which Christian theology needs to (re)assert, Christian theology also highlights the transcendence of God as key to a doctrine of creation. Here it must distinguish itself from Buddhist thought. Ninian Smart has commented on why this notion of God's transcendence is crucial, not only in a theological sense, but a practical one as well:

> But a vital lesson of the very doctrine of creation which lies as a major root of the Christian and other faiths is that the divine Being is transcendent. This itself means that the Christian … is a citizen of two worlds, of two cities – the city of heaven and the city of this world. As such Christianity … has a point of criticism of the world. The practical meaning of God's transcendence lies in part in a readiness to critique the world."[38]

The hurdle for such a radically dualistic ontology, as in Buddhism, lies with its inability to critique the system of which it is a part. There needs to be some connection between the transcendent and finite. Here is where a Christian notion of mediated transcendence can be so vital to critiquing worldviews, governments, anthropologies, etc., whereas Buddhist ontology has little grounds on which to offer an (transcendent) alternative.

Hinduism

In approaching our second major world religion – Hinduism – we are met with Arvind Sharma's position that no strict definition is possible.[39] In fact, Sharma argues that Hinduism is a circumference seeking a center (compare this to Christianity being a circumference radiating from a center – Christ), either in sacred texts, ethnic identity, the priestly class of Brahmins, or some other such entity. Since 95% of Hindus live in India, it is tempting to classify Hinduism as an ethnic religion. However, for Sharma, there are two ways to define a Hindu – as an ethnic Hindu, and as one who believes in Hinduism.

And, as such, "membership of the Hindu community is primarily a matter of self–definition." [40] Swami Vivekānandu is said to have created modern Hinduism in a speech to the World's Parliament of Religions in 1893 in which normative Hinduism is presented as based on the texts called the *Upaniṣads*, or the final sections of the Vedas. Sharma goes on to outline four major contemporary forms of Hinduism. The first is *Absolute Hinduism*, which is the preeminent form of intellectual Hinduism. It posits that humanity is the supreme reality, and that the ultimate state is to discover this by various practices. This has never been very popular. A second more popular form is *Theistic Hinduism*, exemplified by Rāmakrishna, who not only vindicated traditional forms of Hinduism, but also argued for the essential unity of the great religions. The third form is *Activistic Hinduism*, as exemplified by Mahatma Ghandi. This has remained the dominant form of modern Hinduism for most of the century, and posits active service as a path of God–realization, involving nonviolence, courage, faith in God, truth, ecumenism, and self–sacrifice. A subset of activistic Hinduism would be *militant Hinduism*, which advocates a more violent course.

Sharma includes a wonderful analogy that Hinduism is like democracy in the United States. Hindus are free to do as they wish in the spiritual life (there is no established church) as long as it does not transgress on others. Hinduism allows for individual routes to a destination, each involving knowing, feeling and willing. There are four valid goals of existence: 1) moral life; 2) wealth; 3) pleasure; 4) liberation. [41] The task is not to get distracted on the way towards one's goal, while enlisting as many gods as possible along the way.

It is important to note the role of sacred texts in Hinduism. The revelational literature in Hinduism is perhaps the largest in the world. And though Vedic authority is roughly a criterion of Hindu orthodoxy, few Hindus know its contents (Sharma suggests that it is analogous to the U.S. Constitution for American citizens). The Vedas can be seen as *a center*, but they should not be seen as *the core* of Hinduism. In fact, the rise of the importance of the *bhagavadgita* may be seen in the context of interreligious dialogue as a response to Christians being people of the book. Moreover, in terms of the Vedic content, Hinduism is not as interested in historical events being narrated as it is in the theological value that comes from converting history into myth. [42] Hinduism is thus to be seen as both inclusive and encyclopedic, valuing synthesis over the exclusiveness of either Christianity or Islam. In short, Hinduism is more of a way of life than a belief system based on one sacred text. [43]

Key Issues in Hinduism Related to a Christian Doctrine of Creation.

These introductory remarks have underscored an issue that is crucial for analyzing Hinduism's perspectives on cosmology and cosmogony, and that is its incredible diversity of ideas. It is actually impossible to say there is one Hindu perspective on cosmogony. For this reason, our course here will be to look at representatives from each of the four main periods in Hindu history, and then analyze some salient topics that arise from our examination.

In the Vedic period (*ca.* 1500 bc. – 800 bc.), we have a number of different notions, according to H. Jacobi:

> The world, according to Vedic notions, consists of three parts – earth, air, and sky, or heaven. But, when the idea of 'Universe' is to be expressed, the phrase most commonly used is 'heaven and earth'. Both Heaven and Earth are regarded as gods and as the parents of gods (*devaputra*), even although they are said to have been generated by gods. Sometimes one god, – Indra, or Agni, or Rudra, or Soma, sometimes all the gods together, are said to have generated or created heaven and earth, the whole world; and the act of creating is metaphorically expressed as building, sacrificing, or weaving. [44]

In a frequently cited philosophical hymn (*Rigveda* 10:129), we see a number of characteristics, including a sort of creation *ex nihilo* and an absence of divine agency:

> Then even nothingness was not, nor existence.
> There was no air then, nor the heavens beyond it.
> What covered it? Where was it? In whose keeping?
> Was there then cosmic water, in depths unfathomed?
> Then there were neither death nor immortality,
> Nor was there then the torch of night and day.
> The One breathed windlessly and self–sustaining.
> There was that One then, and there was no other.
> At first there was only darkness wrapped in darkness.
> All this was only unillumined water.
> That One which came to be, enclosed in nothing.
> Arose at last, born of the power of heat.
> In the beginning desire descended on it –
> That was the primal seed, born of the mind.
> The sages who have searched their hearts with wisdom
> Know that which is, is kin to that which is not.

> And they have stretched their cord across the void,
> And know what was above, and what below.
> Seminal powers made fertile mighty forces.
> Below was strength, and over it was impulse.
> But, after all, who knows, and who can say
> Whence it all came, and how creation happened?
> The gods themselves are later than creation,
> So who knows truly whence it has arisen?
> Whence all creation had its origin,
> He, whether he fashioned it or whether he did not,
> He, who surveys it all from highest heaven,
> He knows – or maybe even he does not know. [45]

Here we see a focusing of analysis to one supreme He, however this He is not above creation (in fact he came later than creation), and has no power or knowledge of creation. This He is at the mercy of the "fertile mighty forces."

In the period of the Brāhmanas (*ca.* 900 bc – 650 bc), the speculation of the previous era concerning the origin and cosmology of the universe gives way to the imagery of sacrifice. [46] For example: "Having given his self (*āt-man*) to the gods, he [Prajāpati] then brought forth that counterpart (*pratimā*) of himself which is the sacrifice (*yajña*). Therefore they say, 'the sacrifice is Prajāpati," for he brought it forth as a counterpart of himself." [47] Another development , which has similarities to New Testament cosmology, is the Veda as the Word, or *brahman*, being the source of creation:

> This Puruṣa Prajāpati desired, "May I become many, may I reproduce." He exerted himself; he practiced *tapas*. Having exerted himself and practiced *tapas*, he brought forth first of all *brahman*, the threefold knowledge (*trayī vidyā*). It became a foundation (*pratiṣṭhā*) for him. Therefore they say, "*brahman* is the foundation of everything here." Therefore, having studied [the Veda] one is established on a foundation, for this *brahman* is his foundation. Resting on this foundation, he [Prajāpati] practiced *tapas*. He brought forth the waters out of Vāc, [who was] the world. Vāc alone was his; she was sent forth. She pervaded everything here, and because she pervaded (root *āp*) whatever was here, therefore [she is called] water (*ap*); because she covered (root *vṛ*), therefore [she is called] water (*vār*). He desired, "May I reproduce from these waters." He entered the waters together with this threefold knowledge. Thence arose an egg (*āṇḍa*). He came into contact (root *mṛś* + *abhi*) with it. "May it exist, may it exist, may it multiply," thus he spoke (root *bru*). Thence was first brought forth *brahman*, the threefold knowledge. Therefore they say, "*brahman* is the first–born of this all." For even before that Puruṣa, *brahman* was brought forth: it was brought forth as his mouth. [48]

Here the Word, Veda as brahman, serves not only as the foundation of the entire creation, but of the creator itself. [49]

As we move to the Upaniṣadic (*ca.* 800 bc. – 300 bc.) periods, we see the universe as a cosmic man in another famous passage from the *Rigveda*s (10:90):

> A thousand heads had [primal] man,
> A thousand eyes, a thousand feet:
> Encompassing the earth on every side,
> He exceeded it by ten fingers' [breadth].
> [That] Man is this whole universe,
> What was and what is yet to be,
> The Lord of immortality
> Which he outgrows by [eating] food.
> This is the measure of his greatness,
> But greater yet is [primal] Man:
> All beings form a quarter of him,
> Three–quarters are the immortal in heaven.
> With three–quarters Man rose up on high,
> A quarter of him came to be again [down] here
> From this he spreads in all directions,
> Into all that eats and does not eat. [50]

The *Sāmkhya* perspective from this period, which is more of a philosophical companion to the practice of yoga, describes how the universe "evolved from *prakti* (primordial, undifferentiated matter, regarded as feminine) by becoming entangled with *purusa* (individual soul, regarded as masculine)." [51] However, the emphasis here is on the individual soul – i.e. the yogin gradually extracts the soul from *prakrti* – rather than on a delineating a worldview encompassing all things. In this period, we also note the development of the three roles of *Puruṣa*, *Prajāpati*, and *brahman* as three distinct manifestations of the *Brahman–Atman*:

> Although Puruṣa is at times described as a creator principle who engages in the demiurgic act of creation, it is generally in his manifestation as Prajāpati that Puruṣa assumes his role. Prajāpati is identified in the Upaniṣads with that aspect of Puruṣa which brings forth the universe and then enters into his creation as the animating principle of intelligence that endows embodied beings with consciousness. As the immanent principle in creation, Prajāpati is sometimes identified with the indwelling Self, Ātman, and with that aspect of Brahman–Ātman which is associated with time and has parts. However, he more often assumes a subordinate role as the creator god who himself has his source in the supreme reality of Brahman–Ātman. [52]

This progression – from Brahman–Ātman to Prajāpati – is seen in numerous passages, e.g.: "In the beginning this universe was just water. That water produced the true (or the real), *Brahman* is the true. *Brahman* (produced) *Prajāpati* and *Prajāpati* (produced) the gods. Those gods meditated on the real." [53]

The imagery of the cosmic man continues into the *bhakti* (devotional traditions) period (*ca.* 1100 ad. – 1700 ad.), where the universe is the body of the deity, and all beings form part of the body of the deity. [54] In the Purāṇic perspective (in the *bhakti* tradition), we see cosmology being "at the very heart of the Purāṇic vision." [55] Jacobi sees at least seven different ideas in his analysis:

> The mythological elements of cosmogony are mostly adopted from vedic literature, and further developed .... These mythological elements are as follows: (1) the highest godhead, Brahman or Ātman, identified with Nārāyaṇa, Viṣṇu, Ṣambhu, etc., according to the sectarian tendency of the author; (2) the primeval waters or darkness; (3) the Puruṣa or Hiraṇyagarbha, who sprang up therein; (4) the world–egg, which brought forth Brahmā ...; (5) the lotus, from which sprang Brahmā; the lotus itself came forth either from the waters or from the navel of Viṣṇu; (6) the intermediate creators, or mental sons of Brahmā ...; (7) the successive creations and destructions of the world. [56]

The staggering diversity of ideas is evident, including some approaching a mono–deistic conception, a cosmic man, a world–egg, intermediaries, and successive creations and destructions of the world itself.

Theologically speaking, it is difficult to distill the many ideas from these four periods into any singular Hindu vision. This may, in fact, be the most important point of our analysis – the impossibility of identifying *the* Hindu cosmogony. But there are enough repeat characteristics present in the many examples to warrant comment. The first centers on the mode of creation. For Hinduism, we are speaking of a divine activity, because in the background of almost every cosmogony is a deity. The mode of creation may be emanation, or imposition of order upon chaos, or even some sense of fashioning. But a deity is always present. For instance, Sịva, in the *Śvetāṣvatara Upaniṣad*, is portrayed as a personal deity who is the originato[57]r of the cosmos. The text shows how Sịva is the great Lord, and the many gods of the Vedic hymns are shown to be aspects of him. [58] Later, in 9th century Kashmir Ṣaivism, Sịva is the supreme reality, "whose essence is pure consciousness; he is eternal and immutable in his own nature but underlies the universe as *ātman*, immanent through his creative energy (*sākti*), which is often personified as his consort. Creation is simply the self–projection of consciousness which is essentially free but has the choice both to limit itself by projecting other entities or to reabsorb them." [59] Here creation is decidedly monist and immanent, the projection of consciousness, though in terms of the nature of the divine, we do have one creator.

Rāmānuja (1017–1137), on the other hand, in a famous and followed version of creation, seeks to give reality to the world apart from the deity. *Īṣvara*, the deity, is real and independent, but souls and matter (the two constituents of the world) are also real, though completely dependent on *Īṣvara*. The world is thus the body of *Īṣvara*, and is his instrument and part of his self–expression. [60] So that in Rāmānuja, we have a continuance of creation as emanation, but with a further development of a kind of dualism in that the world exists somewhat on its own. However, we still see the world as the deity's body, and merely a part of his self–expression. There is no true autonomy on the part of creation, and as such we are back at a monist understanding. Rāmānuja's position is critiqued by Madhva (c. 1198–1278), who argues that Rāmānuja's deity and the world are too intermingled. Madhva advocates a strong dualism, and claims that

> the Vedas describe Brahman, that is the deity, as full of infinite perfections and so he dismisses as purely figurative any monistic passages in the Upaniṣads. Equally, he considers that Viṣṇu [a deity], individual souls and matter (*prakṛti*) are eternally and entirely distinct. He thus emphatically rejects any suggestion that the world is a mere illusion or emanation of the deity and propounds an absolute dualism, in which he insists on a total distinction between inert matter and conscious spirit. [61]

We can see from these examples the radical diversity of thought within Hinduism concerning creation. But in the main, we are left with creation as some form of monism, either as the self–expression of the deity, or with the world being merely an extension of the deity. The interesting point here is the relatively important role of the deity, in contrast to typical Hindu spirituality, which is more concerned with the practical. The divine, while construed quite differently than in Christianity, and in contrast to the no–God of Buddhism, is key to a Hindu doctrine of creation.

This brings us to a second set of issues concerning creation, and that is the blurring between transcendence and immanence that is typical of Hinduism (though this is an imperfect, two–dimensional way of conceptualizing the matter). We have already noted how in a Hindu doctrine of God, the emphasis falls decidedly on immanence, with a pantheism resulting from this thrust. This immanetalism springs from the monist understanding of Śaṅkara (c. 788 ad –820 ad), generally regarded as the founder of the Advaita ("non–

dual") school, which insists on Brahman as the sole reality and denies any duality. Śaṅkara's position involves an almost Buddhist theory of two levels of truth, in which, in ordinary experience we perceive differences and multiplicity, but Brahman, as pure intelligence, is divested of any forms, and is the one truth, the ultimate reality. Therefore, Brahman's "unqualified" reality is operative. [62] Here we have the curious combination of ineffability with complete interpenetration!

The Christian viewpoint attempts to hold the aseity of God (transcendence) in balance with his involvement in creation (immanence), and any attempt at diminishing one or the other aspect of this balance has generally resulted in the charge of heterodoxy from within the Christian church. This position is rooted in the doctrine of the Trinity, which attempts to hold together the paradox of transcendence and immanence through the person and work of the Father, Son, and Holy Spirit – a unity, yet distinct. Eberhard Jüngel offers a fine explanation of how this paradox can be maintained in his commentary on the work of Karl Barth in his book *The Doctrine of the Trinity* (1976). Jüngel takes Barth's claim that God's inner workings and his outer workings are one and uses it to tackle a problematic implication of Barth's method, namely, if God is self–moved, completely differentiated from His creature, how then can we speak of a God who is relationally involved with His creature? Does not this make God dependent on something outside of Himself and thus bring into question His aseity? [63] Jüngel argues that relationality already exists within the divine being, within the Trinity – *ad intra* – and so God's relationship to the world – *ad extra* – simply corresponds to that root reality. Thus Jüngel solves the dilemma by preserving God's being–in–Himself (and by implication, his transcendence), while allowing him to relate (immanently) to the creature which is "indissolubly subject". [64] Here we once again observe a way the doctrine of the Trinity is so crucial for a Christian doctrine of creation: It helps keep the balance of transcendence and immanence in the divine.

A number of attempts have been made to conceive of the Hindu God(s) as a Trinity. Sharma notes, "God is visualized in three forms, as constituting the trinity (*trimūrti*) of Brahma, Viṣṇu, and Ṣiva, corresponding to the three cosmic functions of emanation (or creation), preservation, and consummation (or destruction)." [65] Brockington, however, does not put much credence in this configuration – "the grouping does not suggest that the deities are in any real sense aspects of one whole.... Its only real purpose is to function as another mechanism for the acceptance of essentially competing sects worshipping different deities within the one umbrella of Hinduism. The only point of similarity to the Trinity is the trivial one of the common occurrence of three in the formulation." [66] As to this configuration's specific connections with the Christian Trinity, there appears to be little common ground in the specific tasks of each deity, particularly with Ṣiva. We might almost understand the three Hindu deities as simply being distinct.

Nonetheless, there have been a number of appropriations of Hindu thought by Christian theologians in their own Trinitarian conceptions. Certainly Raimundo Panikkar's *Trinity in World Religions* (1970), for example, explores the Christian appropriation of Hindu themes for the devotional life. The three ways (or *yogas*) of the *Bhagavad Gita*, knowledge, devotion, and works (*jñana*, *bhakti*, and *karma*), are to be found in all religions. The spirituality based on *karma* (works) corresponds to God the Father; the spirituality based on *bhakti* (devotion), corresponds to God the Son; the Spirituality based on *jñana* (knowledge) corresponds to God the Holy Spirit. [67] Other examples could be cited as well. Frank Whaling notes several examples of Christian appropriation of Hindu thought on the nature of God. The Absolute (Brahman) could be thought of as *Saccidānanda*, which is a union of the Sanskrit words *Sat* (Being), *Cit* (Intelligence), and *Ānanda* (Bliss). In Christian terms, Being then refers to God in his ineffable aspect, Intelligence to Christ as the Logos, and Bliss to the Holy Spirit, the Giver of Joy. Even further, "Being could signify God in his unmanifest aspect, Intelligence could signify God in his incarnate aspect, and Bliss could signify god in his immanent aspect." [68] Another example is the Hindu idea that God as Brahman is both without qualities (*nirguna*) and with qualities (*saguna*) (cf. Buddhist transcendence). As *nirguna*, God is ineffable, and cannot be known as he is – this is transcendence. As *saguna*, God is personal, and does have qualities and form that can be known (i.e. *Iśvara*) – this is immanence. [69] Here we have an attempt to bridge the transcendent/immanent divide so present in Hindu thought, though from a Christian, not a Hindu milieu.

But again we must note that these are Christian attempts at fitting Hindu doctrine into Christian categories, and as such, do not leave the doctrine to stand in its own Hindu context. Hindu doctrine, in this area, is so diverse that a clear connection with the Christian godhead is difficult. The strict monotheism that guides Christianity (and Islam and Judaism, for that matter) is at odds with the Hindu vision, as is the Christian divide between creator and creature. However, *appropriations* of specific Hindu themes for Christian theologians in the context of a doctrine of God, such as we have just cited, are alive with possibilities. In the end, we are left with very different pictures in Christianity and Hinduism of creation, each with its attendant conceptions of God, the universe, and humanity.

One other interesting connection to be mentioned here is that Christian and Hindu thought both agree on the stewardship of creation. Recently, Christian systematic theologians have sought to see the command in Gen. 1:28 to "have dominion" over the earth not in terms of domination, but in terms of stewardship (e.g. Moltmann). The Hindu tradition has seen a much greater continuity between humanity and the rest of creation, primarily because of the cycle of rebirth; there is no intrinsic distinction between humanity and other animals. [70] This sentiment underlies much of Hindu preference for vegetarianism. However, in practical terms, the tendency mentioned above to regard non–reality as the essence of the universe has given Hinduism the reputation of being somewhat world–denying. This perception of Hinduism has parallels with the perception that Christianity has been at odds with concern for the environment. [71] While the Christian idea of humanity possessing the *imago dei*, in contrast to animals and the rest of creation, stands in contrast to the Hindu viewpoint, still Hinduism may have much to teach Christian theology in this area. Both Christianity and Hinduism affirm the value of creation, so perhaps this is an area where Hinduism can strengthen Christian theology.

Islam

In Islam we meet the other world religion (along with Judaism) that has the most similarities with Christianity, primarily because of its monotheistic roots and its emphasis on one inspired text. Islam's monotheistic roots parlay into an even stronger view than Christianity of the transcendence of God. It is clear that the central doctrine of all Islam centers around God in Himself, as well as in His Names and Qualities. [72] A number of Qur'ānic verses attest to this emphasis: "He has certainly power over everything. There is none who can take away the favours He bestows on man; and there is none apart from Him to restore what He has withheld. He is all–mighty and all–wise" (35,1–2)[73]; "He is the first and He the last, the transcendent and the immanent; and He has knowledge of every thing" (57,3); "God: There is no god but He, the living, eternal, self–subsisting, ever sustaining. Neither does somnolence affect him nor sleep. To Him belongs all that is in the heavens and the earth; and who can intercede with Him except by His leave? Known to Him is all that is present before men and what is hidden (in time past and time future), and not even a little of His knowledge can they grasp except what He will. His seat extends over heavens and the earth, and He tires not protecting them: He alone is all high and supreme" (2, 255). This emphasis both on the omnipotence and omniscience of God is matched by an understanding of the unity of God – He cannot be dissolved into parts. The unity of God is asserted against dualism (16, 51), though it is recognized that the God of Jews and Christians is the same as Allah (cf. 29, 46). This is not the "no–God" of Buddhism, or the many gods of Hinduism, but rather the singular, towering figure of Allah. [74]

God's message in Islam is brought to the people in its fullest form by Muhammad, God's emissary. The role of Muhammad the prophet is not the same as Jesus the Christ for Christianity. Muhammad is not divine, but rather a human possessing the most perfect of natures. And though Islam is based primarily on the Absolute, Allah, and not on the messenger, the "love of the Prophet lies at the heart of Islamic piety, for humans can only love God if God loves them, and God only loves the person who loves His Prophet." [75] Further, in Sufism and many schools of Islamic philosophy, Muhammad plays an important role as being identified with the Logos, God's first creation, "which is the ontological principle of creation as well as the archetype of all prophecy" – this is called the "Muhammadan Reality." [76] Even more important than the prophet's life are the prophet's words – the *Hādith* – which constitute a vast corpus of sayings which are important to all schools of Islamic thought. [77]

It is the Qur'ān, however, which is the central repository of God's message to the world. In it is the "central theophany of Islam, the verbatim Word of God revealed to the Prophet by the archangel Gabriel." [78] The role of this one inspired text is quite different than the multitudinous texts for Hinduism and Buddhism – the Qur'ān is the key reference point for Muslims on issues of faith and practice. In this respect, Islam is again similar to Christianity as a religion of "the book", though Muslims treat everything connected to the Qur'ān – its sounds, the very words, the letters, etc. – as sacred.

In terms of the practicality of Islam, the Divine Law, or *Sharī'ah*, is what Muslims consider to be the embodiment of the Will of God, and what God wants them to do to gain happiness in this life and in the hereafter. [79] The Will of God is both the dominant feature in a devout Muslim's life as well as the overwhelming emphasis in the Qur'ān. [80] The *Sharī'ah* divides all acts into five categories: those which are obligatory; those which are recommended; those toward which the divine law is indifferent; those that are reprehensible; and those which are forbidden. The goal of the Muslim life is thus to surrender to the Will of God, and so gain freedom of their own egos from the "stifling confinement of their passionate selves", as Nasr puts it. [81] This external emphasis is counterbalanced in Islam by an internal emphasis, particularly in Sufism, on the inner or esoteric dimension. [82] Meditation and devotion are key elements to this aspect of Islam. Together, the inner and outer aspects to Islamic life form a well–rounded spirituality.

These significant contours of Islam – the transcendence of God, the role of Mohammed, the centrality of the *Qur'ān*, and the significance of the Divine Law – are augmented by the crucial role of wisdom or intelligence for the Islamic system. F. Schuon notes, "the basis of spiritual ascent is that God is pure Spirit and that man resembles Him fundamentally through the intelligence; man goes towards God by means of that which is, in him, most conformable to God – the intellect." [83] According to Schuon, humanity is made up of intelligence and will, and so Islam's anthropology is met by Islam's theology: "If man is will, God is Love; if man is intelligence; God is Truth. If man is will fallen and powerless, God is redeeming Love; if man is intelligence darkened and gone astray, God is the illuminating Truth which delivers; for it is in the nature of knowledge – the bringing of intelligence to the level of truth – to render pure and free."[84] The significance of intelligence and will for anthropology is seen in the soteriological emphasis in Islam on subduing the passions which cloud the will.

Key Issues in Islam Related to a Christian Doctrine of Creation

As we hone in more precisely on the Islamic doctrine of creation, a good place to begin is with the one inspired text – the Qur'ānic data on the word, *khalḳ*, or "creation". *Khalḳ* has a wide range of meanings and encompasses a number of different words. But our initial focus is on the verb *khalaḳa*, which for R. Arnaldez means "to create" in the same sense as the Hebrew *bārā*. [85] Many Qur'ānic verses replicate Genesis 1:1, e.g. "All praise be to God who created the heavens and earth …" (6, 1), and add that God also created whatever is in the heavens or on earth (e.g. 10, 6). As the term is further used for other categories of beings, khalaḳa comes closer to another family of words (*sawwā*, *ṣawwara*, *ḳaddara*, and *ansha'a*, etc.) which seem to indicate more of a sense of "fashioning": "It is He who produced you from a single cell" (6, 98); "It is He who raised you from the earth" (11, 61; cf. 53, 32). This further sense of fashioning includes an idea almost of another creation in the creation of humanity: "Then We fashioned the sperm into an embryo, then fashioned the embryo into a shapeless lump of flesh; then from the lump of flesh We fashioned bones, then clothed the bones with flesh. Thus We formed him into a new creation" (23, 14; cf. 29, 20). Every act of God is seen as another act of creation. It then follows that we should see a sense of *creatio ex nihilo* in this doctrine, e.g. "Creator of the heavens and the earth from nothingness, He has only to say when He wills a thing: "Be", and it is" (2, 117; cf. 3, 47). Arnaldez further notes,

> Thus the creative Word, the *kun* (equivalent to the biblical *fiat*) is not a *flatus vocis*: it brings about being in actual truth. Creation has an end, although God has no need of anything apart from Himself: He is sufficient unto Himself .... Creation, say the theologians, disagreeing with some conclusions of Greek philosophy, does not emanate from Him by a natural and necessary process.[86]

Here there is no pantheism, or emanation, or development of creation: only divine freedom in totality. Summarizing, in terms of cosmogony, "it is possible to speak of a creation *ex nihilo* which is primordial and universal, then of succeeding creations which give life to particular beings of which occur at different stages of their development."[87]

This primordial sense of creation (*khalḳ*) is expanded in the Qur'ān to include further aspects of creation. The purpose of the work of creation – "I have not created the *jinns* and men but to worship Me" (51, 56) – ultimately ends in a return to God: "God originates creation, and then will revert it, then you will go back to Him" (30, 11). Humanity is seen as God's viceroy on earth (cf. 2, 30; 31, 20), stewarding the earth, and is therefore "the purpose of creation, because he has been created with the power to know God, to obey Him, to worship Him, and to return to Him."[88] Eschatology is drawn (similarly to the Christian vision) into the doctrine of creation as well – the creation must have an end (cf. 6, 2).

Further amplifications of an Islamic doctrine of creation are brought to life by their theologians. "Generally [the Qur'ānic commentaries] state precisely that it was a creation *ex nihilo* and often explain *khalḳ* by *īdjād*, the gift of existence."[89] With other theologians, especially the Mu'tazilīs, the idea of preservation and continuous creation begins to emerge – cf. "Who creates first then reverts it? And who gives you provision from the heavens and the earth? Is there any other god along with God?" (37, 64). This sustaining power is associated with creating power in the oneness of God – creation must be repeated at each moment of existence, or else creatures would slip into the void. [90] This type of analysis pushed theologians to begin analyzing the nature of God – i.e. was God eternally "creatory" or simply, before the creation of the world, "non–creatory"? The Ash´arīs believe that "creatory" as a name of God was eternally applied to him. [91] If theologians were speculating as to the eternality of God, then the question of the eternality of creation was bound to arise as well. To assert that creation is eternal, however, is to fall into the trap of asserting something co–existing with the Creator, which impinges on the Creator's sovereignty. [92] To assert that creation is not eternal, and is in time, pushes one to examine the motives and will of the Creator. Did he have ideas (i.e. forms) in his mind that existed prior to the actual created thing? And this tack resulted in even further emphasis on the divine freedom and transcendence of God among the Mu´tazilīs. For example, in Islam there is no intermediary in creation (*contra* Christianity, where the Christological mediation of creation is crucial). There is neither anything that resembles God (cf. 42, 2). This has strong implications for anthropology, especially regarding the *imago dei*, an important concept for Christianity.

In terms of some of the specifics of creation, the Qur'ān sets forth a chronology:

> Do you refuse to believe in Him who created the earth in two spans of time, and set up compeers to Him, the Lord of all the worlds? He placed firm stabilizers rising above its surface, blessed it with plenty and growth, and ingrained the means of growing its food within it, sufficient for all seekers, in four spans. Then He turned to the heavens, and it was smoke. So He said to it and the earth: "Come with willing obedience or perforce." They said: "We come willingly." Then He created several skies in two spans, and ingrained in each sky its function, decking the nearest heaven with lamps, and guarded it. This has been determined by the mighty and all–knowing. (41, 8–12)

This text seems to indicate some sort of fashioning, not creation *ex nihilo*; but of course this must be balanced against the other many texts supporting the power and singularity of the divine will. The specifics of cosmogony and cosmology are less attested to than is the divine being.

One more interesting note regarding the specific days of creation is in order, and that concerns the Sabbath. The Sabbath is a crucial concept in a

Christian (and Jewish) doctrine of creation, and for some theologians (notably Moltmann), provides the essential framework for dialogue and mutuality in pluralism. However, in Islam, especially in the *hadīths*, many narratives about the days of creation are designed to refute the Jewish concept of the Sabbath day of rest (cf. 50, 38): God never rests because tiredness never affects Him. [93] Moltmann's model of divine participation in life – the Creator enjoying creation, resting, promoting fellowship and communion among creation; in short, the immanence of God infusing all life with joy and rejuvenation – does not seem to connect at all with the Muslim vision of transcendence and law. [94]

As is apparent from this look into the lexicographical data, many of the themes which we outlined in our broad look at Islam are a part of its doctrine of creation as well. In terms of transcendence and immanence, the emphasis thus far has been on transcendence. However, this does not exclude in Islamic thought any sense of divine immanence. One of the names of God is "the Near" (*al Qurīb*) (though it does not appear as one of the ninety–nine names in the *Tirmidhi* list). [95] The Qur′an also indicates that God is closer than one's jugular vein (50,15). And certainly Sufism has tended to emphasize the more immanent aspects of God in interiority, devotion, meditation, etc. Nonetheless, Islam has no incarnation, as Christianity does, nor does it have the interpenetration so characteristic of Buddhism, or the reincarnation of Hinduism. Even the attributes of God that would tend towards a more immanent understanding – compassionate, merciful, forgiving, etc. – must be understood as the prerogatives of power, and not as an attempt by God to become closer to humanity. [96] These attributes are not to be seen as any different than the essence of God, though some (particularly *Abu–Hudhayl*) see the attributes as modes of the essence of God. [97] Seen through this prism, the attributes of God could be compared to the persons of the Trinity in Christian thought, particularly in the work of Barth, who uses the word "mode" to discuss the Trinity. Though for both Barth and Islam, there would be no question of the ineffability and unity of the divine subject; both are merely the describing of the practical work of God.

There are other issues related to a doctrine of creation that we have hinted at, but not developed fully. One is that of continual creation, which is analogous to the Christian doctrine of providence. We have already mentioned the sense of creation for each successive act in Islam, and this is the way Allah maintains and sustains the universe. W. Montgomery Watt notes that "more generally, the Qur'ān emphasizes to a much greater extent than the Bible the fact that God constantly controls all mundane events, both natural and historical."[98] The Islamic doctrine of continual creation has similarities with the Christian sense of the Spirit of God upholding Creation in a providential sense. A main issue here (as it is in Christianity with Calvinism) is what place does humanity hold then in the scheme of the world? [99] Are humans free to do as they please, or are they merely puppets of God? Predestination has been a hotly contested issue in Islam, and space does not permit a full briefing here. But in general, the overwhelming Qur'ānic emphasis on the divine will and power mitigates against seeing humanity as possessing free will. [100]

A second issue that must be explored for a full–orbed Islamic doctrine of creation is its connection with anthropology. Though a strong divine transcendence is present, alongside of this is something correlating to the Christian doctrine of the *imago dei*, a doctrine that is absent in Hinduism and Buddhism. The Qur'ān states, "I have fashioned [man] and breathed into him of My Spirit" (15, 29). Nasr writes, "Men and women are created in the 'image' (*ṣūrah*) of God, according to the famous Prophetic *Hadīth*. Here *ṣūrah* means the reflection of God's Names and Qualities, for otherwise God is formless and imageless."[101] And similar to Christian theologians, Islamic theologians have struggled to ascertain exactly what is meant by this Spirit or Image in humanity. But clearly, in Islam the transcendence of God cannot be impinged upon or even correlated in humanity. Also similar to Christianity is the "fall" of humanity, which the Qur'ān attests to. But in this "fall", humanity did not commit an "original sin" which would radically distort human nature, but rather is tainted with the "forgetfulness" which characterizes fallen human beings. [102] Thus, the Qur´an can state "If God were to punish men for their iniquity He would not leave a single moving thing on earth" (16, 61). However in general we can state in Islam the diagnosis for the human condition is not seen as pessimistic as the Christian diagnosis is, so the cure does not need to be as radical either.[103] For example, continuing the scripture quoted above (16, 61): "Yet, He gives them latitude for a time ordained. When that time is come, there will not be a moment's delay nor a moment's haste." The road to "redemption" then runs primarily through intelligence and will, which need to be guided by revelation. Nasr writes eloquently of this process for the human:

> The function of religion [in Islam] is to remove the veil of the passions, which presents the intelligence from functioning correctly. Religion is essentially the means for men and women to recollect who they are and to return to their inner and primordial natures that they still carry within themselves. Humans must be perfectly passive toward heaven, hence the servant or slave of God (*'abd Allāh*), and active toward the world around them as God's viceregents on earth. [104]

At this point we cannot delve further into the topics of salvation or sanctification; our concern is for the state of the human as created. Another angle on humanity is presented by R. Arnaldez in his article on *insān* (man). He notes that God created man weak (4, 28); and that in trouble he cries to God, and when the trouble has passed, he forgets (cf. 10, 12); he is very unjust (14, 34), and a subtle reasoner given to argument (18, 54).[105] It seems that like the Christian human, the Muslim human is in conflict with his created nature in the reality of his or her everyday decisions. [106]

Points of Connection & Disconnection Among the Four Major Traditions

It will be helpful at this point to summarize some points of connection and disconnection between these four major religious traditions within the Christian doctrine of creation. We are using the terms connection and disconnection to connote not merely similarities or differences, but areas of potential dialogue. Our goal in a pluralistic inquiry is not to gloss over the presence of real differences for the sake of a homogenous "worldview" that all four religions can adopt, but rather to let each tradition speak for itself, and then draw connections and disconnections from that exercise.

In general, we must say that with regard to the Christian doctrine of creation there are far more differences between the four traditions than there are similarities. In fact, the areas of connection among all four are few, though they may provide for us a good beginning point for dialogue and even insight. But, before we begin this summary, we must comment on the nature of agreements and disagreements in comparing religious thought. It is clear that one could focus solely on the connective points – that is, areas where the different traditions agree – and build a pluralistic theology on those points. It is also clear that learning from the "other" may involve learning from points of disagreement. That is to say, Christian theology may need correction by the Spirit speaking through other religious traditions. This has been our contention, that a pluralistic theology rooted in a pneumatological natural theology will recognize and incorporate insight from other religions as a helpful resource for Christian theology. That insight may come in the form of similarity to the Christian vision, or it may come in the form of a challenge to the Christian vision. But either will be welcomed as a resource for theology. As a result, this dialectical section will note both areas of agreement and disagreement, but then will then hone in on three that may hold the most promise for fruitful integration into Christian theology.

1. Nature of God – Here the differences between the traditions are readily apparent and well known. There is no supreme being in Buddhism, and the figure of the Buddha plays no significant role in creation. The other three traditions – Hinduism, Islam, and Christianity – all outline a crucial role for the divine. Hinduism posits many gods, and the divine presence is key to any cosmological understanding. There are even attempts at distilling the Hindu pantheon into a trinity of sorts. Islam, in contrast, puts forth one supreme, powerful God, alone in transcendence, creating and preserving the world. This Islamic perspective is shared to a degree by Christianity, which argues for one God, creating and preserving the world. We have noted the Christian tradition's ability to harmonize Hindu thought with Trinitarian theology. We have also noted how Christian theology can learn from its seemingly distinct partner in Buddhism in the idea of not predicating attributes to God. Buddhism's ready admittance of a lack of knowledge of the divine could be a helpful corrective for Christian theology, and some Christian theologians have already been leading us in this direction. A further area of exploration may be the "modes" of the divine as a helpful hermeneutical tool for consolidating agreements regarding the divine essence. But again, by staying within our tradition, and articulating it while being informed by other religions, we believe we are giving not only our tradition a full hearing, but the others as well. Interestingly, while there are major disagreements as to the nature of the divine, each religion holds a significant place for transcendence. This may be an area to consider in building a pluralistic Christian doctrine of creation.

2. Nature of Humanity – Here again we encounter significant differences between the Buddhist system and the others. There is no self or soul in Buddhism; the ultimate goal is to be dissolved into emptiness in nirvana. In light of the lack of a "salvation" for a soul, the emphasis is not on orthodoxy, but orthopraxis, or moral excellence. In Hinduism, the soul passes to another upon death, and so is reincarnated. However, there is no special *imago dei* that differentiates humans from, say, animals, in light of this linkage with all other parts of creation. In Islam and Christianity, we encounter two versions of the *imago dei* concept. While Islam does not have the fully developed conception that Christianity has (and in fact argues against anything approximating the divine image), it does locate the relatability of the human with the divine primarily in the areas of intelligence and will. If human passions are subdued, and intelligence and will enabled, then the human will succeed in pleasing God. Islam's high stress on the divine intelligence and wisdom, though, has opened it up to charges of Platonism. As we have observed, by setting up another principle (i.e. logos) outside of God, Islam struggles to relate the divine to human in a way that Christian theology does not because of its theology of mediation. The Christian viewpoint is that humanity is created in the image of God, is fallen, and is redeemed in Christ. The soul lives forever and is representative of one person, though Christianity attempts to limit the Greek soul/body dichotomy.

However, there are still possibilities for building pluralistic theology here. There are points of connection with all four religions in the need to do right, or be good. A pluralistic theology of sanctification or ethics holds great promise, in our judgement. Moreover, Christian theology, while adhering strongly to the *imago dei* for humanity, may need to learn from Hinduism's embedding of humanity within all of creation. We will have more to say about this in our discussion of the stewardship of the cosmos as one of our areas to pursue further.

3. *Dualism* – Both Buddhism and parts of Hinduism posit a two–tiered ontology: there is the world that is real, and there is the world that is illusory. We are living in the world that is illusory, and the goal is to get to the "real" world (though for Buddhism that would be "nothingness"). Christianity's insistence on creation *ex nihilo* (matched to a degree by Islam) is an attempt to mitigate against dualism ontologically, as is its belief in existence. The human does exist in reality for Christianity, and not in an illusory world. Also significant here is the role of the divine in the world that we live in. Buddhism and parts of Hinduism suggest the divine or ultimate is beyond our existence, which Christianity can agree with in some form. However, Christianity insists on a means of accessing that divine through a theology of mediation. Other Hindu perspectives note that the world is seen as an exten-

sion of the deity, and so we have a completely immanent understanding of the divine. Islam, of course, will have nothing to do with a dualistic ontology of existence. God is all–supreme and the world is his creation, though it seems to be fashioning of some sort (implying materials already present outside of Him). The challenge for both Islam and Christianity, with their strong monotheism, is to avoid falling into monist categories – i.e. that the world is simply God and nothing else. This is a challenge for parts of Hinduism as well: If the world is merely an extension of the deity, is there an ontological freedom for creation? Christianity seems to do a better job of giving the world its own created existence apart from God than does Islam with its overarching, towering will of Allah. [107]

Given these areas of disagreements, what can Christian theology learn from the "other"? Christian theology has insisted on a lack of dualism at the beginning of creation – creation from nothing – but has also insisted that the world contains its own existence. It is not the extension of a deity. As such, Christianity is in the unique position of both affirming present existence, while at the same time affirming its belief that the beyond, eschatologically, holds the ultimate value. Perhaps a pluralistic Christian theology can build on the insights of Christian thinking in the importance of the here and now, while incorporating more Buddhist insights on the divine being beyond our grasp, while still being the ultimate object of desire. In this vein, we would also note that the importance of the here and now in Christian theology is built, in part, on the freedom of creation, which we will have more to say about in chapter seven.

4. Cosmology – The cosmology of Buddhism is one of three broad categories – the single–world system, with its cosmos as a flat disk, a cosmology of "thousands", with thousands, even millions of worlds, and a cosmology centered around "Buddha–fields". Cosmology in Hinduism is staggeringly diverse, from the universe as world–egg, to the world as the body of a deity, to the universe as the emanation of a personal deity. Islam is similar to (built upon?) the Christian and Judaic perspective. The Christian viewpoint is certainly Ptolemaic, with the earth central, surrounded by the watery abyss. Here we will quote from Genesis 1:

> In the beginning when God created the heavens and the earth, the earth was a formless void and darkness covered the face of the deep, while a wind from God swept over the face of the waters. Then God said, "let there be light"; and there was light…And God said, "Let there be a dome in the midst of the waters, and let it separate the waters from the waters." So God made the dome and separated the

> waters that were under the dome from the waters that were above the dome ...And God said, "let the waters under the sky be gathered together into one place, and let the dry land appear." And it was so. God called the dry land Earth, and the waters that were gathered together he called Seas...And God said, "Let there be lights in the dome of the sky to separate the day from the night; and let them be for signs and for seasons and for days and year, and let them be lights in the dome of the sky to give light upon the earth." (Gen. 1:1–3,6–7,9–10,14–15 – NRSV)

Above the earth is the firmament, which supports the heavens, which contain the fixed stars and the hosts of heaven. Sheol, or Hades, is beneath the earth, and is the waiting–place for departed spirits. [108] These cosmologies exhibit striking differences, but of crucial concern here is what the cosmologies indicate about the role of humanity, the type of God, the form of creation, etc. Is the universe earth–centered? Can Buddhism's insistence on the infinity of the universe be a corrective to an earth–centered Christian capitalism? Where is the center of the world? Both Budhhism, with its recognition of Mount Meru, and Hinduism, with its description of the world as the deity's body, seem to particularize and recast a cosmological story in local terms. Can the Christian story be more conducive to incorporating all of humanity into creation's purposes? Of particular note here would be the hermeneutical issues involved in adapting ancient cosmologies to a modern scientific viewpoint.

5. *Cosmogony* – Buddhism has essentially no doctrine of creation, per se. Its central concept in this area is the doctrine of "dependent co–origination", by which everything arises together. This doctrine is key for Buddhist ontology as well in that everything in the world depends on something else for its existence. Lacking in this conception is a divine agent. Though there are many different cosmogonies in Hinduism, creation by a personal deity is a part of the Hindu worldview, but it is not creation *ex nihilo* – it is emanation or fashioning of some sort. Either of these two options imply a monism and dualism, respectively, that is against the Christian understanding. Islam and Christianity both imply creation by divine fiat or word, coming from the power and ability of the transcendent God. But again, we have noted the important role of the other, the community, in defining who we are, springing out of attention to the Buddhist worldview.

6. *Eschatology* – We mention this area simply because it is a key part of the Christian viewpoint. A full–fledged doctrine of creation will include the "new creation" of the New Testament. Islam is similar to Christianity in that its creation has an endpoint (cf. 6, 2), a *telos*, from a designer. Buddhism's *telos* fades into the nothingness of nirvana, with no divine agent, and no ap-

parent connection with the beginning. Hinduism's future is cyclical, with little or no *telos* in reincarnation. Some of Hinduism's cosmogonies even include successive creations and destructions of the cosmos. The question arises here, what are the consequences of a lack of an eschatology? If Christianity is seen to have a *telos*, would that make it more or less concerned about the status of the world? If Hinduism does not have a *telos*, does its here and now aquire more or less significance? One could certainly make the case (along with Feuerbach) that an other–wordly focus leads to a disinterest in the present state of world affairs. Significant for eschatology here will be not only the "end," but what the end imbues the present with.

7. *Immanence* – The differences here in this subset of the transcendence/immanence continuum are interesting because we will find agreement on the transcendent side. On the immanent side, two out of the four traditions include a place for immanent participation of the divine in human life. Buddhism has no divinity, so there is no place for divine immanence in its ontology. Buddhism replaces divine involvement with a radical interconnectedness – everything participates and in fact defines the existence of the other (though we have also made the point that this connectedness has emptiness at its core, and as such points toward a radical transcendence instead). Many of Hinduism's cosmogonies are noticeably immanent in terms of the divine – the universe as the deity's body, the universe as a necessary emanation and development of the deity, etc. There are also certain types of transcendent divine figures, e.g. in the hymn quoted from the *Rigveda,* which display an impotence in the face of the forces of nature. Christianity attempts to maintain a place for the immanent participation of the divine. Specifically regarding creation, in the concept of the Sabbath we see the divine participation in human existence. Christian theology also endeavors to keep the balance between immanence and transcendence in its divine life, primarily through its doctrine of the Trinity and the person of Christ. It is in "orthodox" (Sunni) Islam where we encounter the weakest immanent conception of God, though the Qur′an has a touch of it (cf. 50,15), and Sufi'sm emphasizes that aspect in more depth. Allah is omnipotent and omniscient reigning ineffably over the earth, and his will is supreme. Humanity is to serve God, and the Qur'ān emphasizes heavily the distance between humans and God.

Focusing the Dialogue Among the Four Major Traditions

Although differences outweigh similarities, this foray into interreligious dialogue has suggested several areas worthy of further exploration. In particular, we would note three.

1. Preservation – Though certainly not all four traditions would use this term, it does convey the sense of a providential upholding of the universe that is present in each system. In some cases, this upholding is the work of a deity; in others, e.g. Buddhism, this upholding, or preservation comes through interconnectedness. Buddhism, without a divine agent, uses its doctrine of interconnectedness to hold the universe and humanity together. Everything and everyone is contingent upon each other. Where this sense of preservation struggles conceptually, though, is that the goal of the devout Buddhist is to escape everything and everyone in a more or less solitary search for *sunyata*, or emptiness. So that while everything in the cosmos is contingent, the ultimate end is to leave that contingency, and so it would seem that Buddhism is working against its own ontology. [109] In Hinduism the concept of preservation is seen in the deity Viṣṇu, whose job description has broadly been described as preservation. Though the many deities and cosmonogies of Hinduism mitigate against one singular understanding of preservation, it is clear for the average Hindu, the divine plays a large role in getting to whatever goal they may have for their life.

As we move into Islam and Christianity we see how preservation links to a concept of creation. For Islam, especially the Mu´tazilīs, each successive act of God is seen as a secondary creation – this is the way Allah maintains the universe. God is in control of all events, sublime and mundane, through his creative will. Compare this to the Christian perspective, which shows God in control of the world, both at the beginning in creation, and continually, through the doctrine of providence. Key for the Christian concept here is the person of the Spirit, who provides both the connection and individuation of human life. We see in all four traditions the concept of preserving or upholding the universe. This may be a helpful place to begin dialogue in terms of cosmogony in a pluralistic context.

2. Transcendence – Interestingly, all four religions hold a place for transcendence. In Buddhism, nirvana is the transcendent goal, but as befits the ontology of Buddhism, it is not to be described at all. Human terms will distort its true nature. Transcendence then is a category for Buddhism, without the substructure of divine life to support it. Hinduism highlights a key role for the divine in almost all of its cosmogonies, particularly in the deity of Brahma in the Advaita school. In the Brahma is to be found true reality: The human is to align him or herself with that ultimate reality. In Islam, we have already noted the transcendent nature of Allah, with the entire worldview being grounded around that pole. Christianity, too, affirms a high place for the transcendent God – e.g. "So acknowledge today and take to heart that the Lord is God in heaven above and on the earth beneath; there is no other" (Deut. 4:39). His transcendence must be maintained, though balanced with a sense of immanence. So it seems that another helpful starting place for dialogue in this context would be a discussion of the "other" – that which is transcendently different – which we aspire to.

3. *Stewardship of the Cosmos* – And finally in the area of taking care of and living with creation – the world – do we find points of contact. We have already mentioned the significance of preservation for the four religions; stewardship of the cosmos is the specifically human task in that area of preservation. In Buddhism, the emphasis on orthopraxy, the here and now, and even a non–anthropocentric view of the universe lends itself to living in harmony with the created order. Certainly the ontology of interconnectedness encourages and values cooperation on the human, animal, and environmental level. This sense of contingency in anthropology could be a key starting point for dialogue with the Buddhist tradition. Hinduism's doctrine of reincarnation also reinforces the continuity between humanity and the rest of creation. A devout Hindu would be careful to care for all the creatures of the earth. In Islam, humanity's role on earth is to be God's "vice–regent", but this role is to be one of stewardship, of caring for the earth in a non–domineering manner. Christianity as well is appropriating a more gentle understanding of "having dominion over the earth" (Gen. 1:28), though of course we must mention Christianity's difficulties in this area. Christianity, along with the other world religions, certainly has a mixed record on this point. Taking care of the earth, and living in harmony with the earth and the rest of creation are viewpoints beginning to seep their way into current Christian doctrines of creation. But Christian theology can perhaps have much to learn from other traditions here. A key beginning point for dialogue here would be, "how do our traditions encourage us to live together and preserve the earth?"

Conclusion

It is clear that there are not as many points of connection in the area of creation and cosmogony as one might suspect. However, three areas of agreement or confluence do provide us with hopeful starting points for dialogue, learning, and incorporation. The more crucial question pertains to the outcome of our task: if we are open to information from the other, what will we do with that information? Will we set it aside and let our theologies of religion deal with integrating it into our theological system? Or, will we let it guide and inform our very theologies? Here our development of the importance of general revelation and natural theology pays dividends: if God is in fact working in the wider world, even through other religions, we must pay attention to that revelation. And in fact that is what we will do in chapter seven, through incorporation, what Panikkar calls "interpenetration," [110] without homogenization. This can occur through a re–evaluation of (in this case) Christian doctrines in light of the viewpoints of other faiths, not a modern re–visioning of religion for all traditions. [111] Gunton and Smart have provided for us two differing models of openness to other traditions. Gunton's self–contained Christian model did well with special revelation, but faltered in incorporating general revelation. Smart's model did well at incorporating general revelation, but was too vague in terms of being true to the Christian tradition. Our goal will be to produce a more adequate Christian doctrine – one rooted in the Christian tradition, but open to general revelation – in the area of creation. [112] In terms of our own exploration, then, we will focus on the three areas – preservation, transcendence, and stewardship of the cosmos – where we have observed broad commonalities within the four major traditions.

CHAPTER SEVEN

A Pluralistic Christian Doctrine of Creation

We must now return to where we began, the Doctrine of Creation, to elaborate on those aspects of a doctrine of creation that might be informed by other religions. In our final chapter we will focus constructively on three areas that are central to a Christian doctrine of creation, and that can be addressed as well by voices from outside the Christian tradition. The three areas that we will be pursuing are preservation, transcendence, and stewardship of the cosmos. In our previous chapter we identified these three areas as ones that hold special promise for pluralistic Christian theology in the area of creation. This chapter, thus, will provide for us an example of how to do the type of pluralistic theology we are advocating, and will be the fruition of the model we have been developing thus far in this book.

Preservation

From the Christian point of view, preservation is a part of the doctrine of providence, that doctrine that emphasizes the control that God has in governing the universe. Such a theocentric viewpoint presents challenges when we are attempting to learn from other systems that may not be so theocentric. Nevertheless, in our earlier analysis we discovered a preliminary area of similarity in the area of preservation. Let us now return to the specifics of those findings in other religions.

Buddhist cosmology is centered around the concept of "dependent co–origination", which is not only a non–theistic cosmogony, but also a non–

theistic cosmology. Dependent co–origination has at its core the notion of "emptiness", that is, that no person or thing possesses existence on its own, but is rather contingent upon everything else. This radical dependence in ontology is mitigated somewhat by the understanding of *nirvana*, which has as its goal the complete self–emptying, the dissolution of all things. Moreover, the third aspect of the Triple Refuge – "I go for refuge to the Community (*Saṅgha*)", typically involves more discussion about the monastic community than the real connectedness of the average adherent. Nonetheless, in Buddhism's ontology lies a strong case for the role of the other in helping to define and preserve who we are as humanity. [1]

We have referenced as well in Hinduism the key role of the god Viṣṇu, perhaps the most popular god in Hinduism. In Viṣṇu, and contemporary Vaiṣṇavism, the preserving aspects of a divine deity are similar in many respects to a monotheistic understanding of preservation. Viṣṇu is considered to be immanent in all things as well as transcendent – the universe owes its existence to him, and pious Hindus accept him as their only and supreme Lord. Klostermaier notes, "The most popular, and consequently the most important part of Viṣṇu mythology centers around the avatāras, the bodily descents of Viṣṇu exercising his function as savior of the world. The Bhagavadgītā explains that the Supreme One comes down to earth whenever dharma is in danger, to save the good and to destroy the wicked." [2] This conception is similar to monotheistic ones, and Klostermaier continues with a short sketch of Viṣṇu theology:

> Īśvara, the Lord Viṣṇu. is the ruler of all, the giver of all gifts, the preserver of all, the cause of all effects, and he has everything, except himself and his own consciousness, as his body. Īśvara is thus the material cause of the universe, becoming through the act of creation the efficient cause, as well as the concomitant cause of all things, insofar as he is immanent in time. Īśvara, who animates the whole world, is not touched by its imperfections. He is omnipresent through his form, through his knowledge of all things and events, and through his body. He is *sat–cit–ānanda*, being consciousness, and bliss, and he is free from sin. He is the refuge and the protector of all beings, having ch qualities of gentleness and mercy. [3]

Here we see the key role of immanence, even pantheism, in preservation for Hinduism. Viṣṇu is in all things, and as such the role of humans is to prepare themselves for God's grace through self–surrender. Rāmānuja's words give the way to salvation:

> The pathway through which the supreme Brahman is to be attained is as follows: By an accumulation of great merit the sins of the past lives are destroyed. Thus

> liberated a person will seek refuge at the feet of the Purusottama. Such self–surrender begets an inclination toward Him. Then the aspirant acquires knowledge of reality from the scriptures aided by the instruction of holy teachers. Then by a steady effort the *bhakta* develops in an ever increasing measure the qualities of soul like the control of mind, sense–control, austerity, purity, forgiveness, sincerity, fearlessness, mercy and non–violence. The *bhaktas* continue with the ritual duties and offer their very own self at the lotus–like feet of the Puruṣottama. They ceaselessly worship Him with dedication. The Puruṣottama, who is overflowing with compassion, being pleased with such love, showers His grace on the aspirant, which destroys all inner darkness. In such a devotee there develops *bhakti*, which is valued for its own sake, which is uninterrupted, an absolute delight in itself, and which is meditation that has taken on the character of the most vivid and immediate vision. Through such *bhakti* is the Supreme attained.[4]

What we see here is the combination of a pantheistic deity with a corresponding individualistic salvation. With Viṣṇu in all, the way to find salvation is to nevertheless worship the deity and devote oneself to him. There does not seem to be a corporate understanding of salvation. In Buddhism, the lack of a divine agent has pushed a corporate human–connected ontology, whereas in Hinduism, the presence of a divine agent has stunted a communal understanding of preservation.

Islam's sense of preservation is captured in the doctrine of continuous creation, whereby God repeats creation at each moment of existence to keep creation from disintegrating. "Every day He is an act" (55, 29) states the Qur'ān, and this highlights the understanding that continuous creation is not so much a process as it is a succession of acts. This continuous creation is seen as impacting creation, but not affecting any development in the divine, as Islam is not wanting to impinge on the creator's sovereignty: "Creation thus envisaged is not a process, but an instantaneous act; and the changes which natural and human history imply are the successive revelations of that act in time, which, along with extension, number, and the other conditions of existence, exist for man, but not for God."[5]

What then can we take from these sources to bring to a Christian doctrine of preservation? Following our model outlined at the end of chapter five, after discovering source material in other religions, our next step is to locate similarities and/or dissimilarities to a Christian understanding. In Buddhism we see the role of all other things in establishing our existence. This thrust has some interesting similarities with the Christian idea of community, of *koinonia*. Obviously the challenge here is that in Christian thought there is God holding everything and everyone together, and the lack of a divine agent in Buddhism is at odds with the typical Christian understanding of preservation. Nevertheless Buddhism has pointed us in a direction with our doctrine of preservation that we must explore: the role of the other, the role of community, in the preserving work of God. Even though Hinduism's emphasis in Viṣṇu is his presence in everything, a pantheistic view, a corresponding emphasis on community, or the other, is not seen. Here we have a perspective somewhat similar to the Christian one, in that the divine is involved intimately with creation, but without a corresponding view of the importance of the other. Typically Christian theology at this point will emphasize the governance or control of God in preservation. There has not been much stress on the role of the community in the sustaining work of God in preservation. Hinduism, with its God (at least in Vaiṣṇavism) present in everything and everyone, also lacks the tenor of overall control on the part of the divine, which is central in Christian thought. Islam's accentuation on God successively creating each new act reminds us of the scripture passage, "The steadfast love of the Lord never ceases, his mercies never come to an end; they are new every morning; great is your faithfulness" (Lam. 3:22–23). Have we unconsciously crept into a deist understanding of the world, whereby God has done his original creative work, and now sits on the sidelines, letting the world evolve as it may? Perhaps an emphasis on the continuous daily creative work of God needs to be highlighted in Christian thinking on preservation.

With these "connecting" points as the sinews of our nascent statement on preservation, we must now return to the backbones, so to speak, of our doctrine in Christian thought. Following our model, what are some guideposts present in the Christian tradition in this area? To begin with we note that providence has been a key aspect to Christian doctrine, and even a doctrine of creation: Calvin states, "To make God a momentary Creator, who once for all finished his work, would be cold and barren." [6] Indeed, to state that God began his work, but has left us to complete it, goes against the conviction that God is for "me", is "present", and will continue to work in the future; this is why theologians have emphasized that "to deny providence is to deny God."[7]

Just how providence is related to creation is seen in the differing terms, "continued creation", and "continuing creation". Weber describes for us "continued creation":

> This means something different depending upon whether one is thinking of "active creation" (the divine act of creating) or "passive creation" (the creatures being

> created and their createdness). In the former instance, the dominant idea is that the creaturely reality always needs the activity of its Creator. It exists only because and to the degree that its God grants it constantly its existence. In the latter instance, it is an inherent quality of the creature to continue to exist because of the unique act of creation which summoned it into existence. Only in the former case does the creature continually require the Creator.[8]

"Continuing creation" emphasizes the here and now, and the Creator is seen as the originator of all that is. Weber notes that it is easy in this approach to dismiss talk of a "beginning", and see God rather as a "first cause", a situation that for him is not a problem given that in orthodoxy, God is also the "first principle", the Being of all being.[9] These distinctions will be important for us as we attempt to incorporate material from other religions into our statement of Christian doctrine.

To further delineate the doctrine, we must outline the oft–stated division of the doctrine of providence into "conservation" (or "sustenance"), "concurrence" (or "cooperation"), and "governance". Conservation, or sustenance or preservation, refers to the maintenance of creation, whereas governance typically highlights the purpose or end to which God leads all things.[10] Concurrence expresses the cooperation of the creature, keeping its independence in terms of will and activity, with the working of God in ruling the world. The assumption here is that the creature is dependent on God, but that "God, nevertheless, maintains things in such a state, and works in them in such a way, that they themselves cooperate as second causes."[11] Berkouwer, for one, is unsure whether this third category is even warranted alongside of the first two, but nevertheless these have been the main categories of the doctrine of providence for orthodoxy. While not denying the interconnectedness of the three areas, we will focus on the area of sustenance, or conservation or preservation, as that area which most relates to the thought we find in other religions.[12] We must also note that of the three areas, Christian theology has tended to focus on government, so that from the outset our honing in on preservation represents a bit of a shift in emphasis, and perhaps a needed one.

As we begin to attempt to integrate our findings from the world religions, we must also take stock of Karl Barth's statement that faith in providence is not "a kind of forecourt, or common foundation, on which the belief of the Christian church may meet with other concepts of the relationship of what is called 'God' with what is called 'world.'"[13] Elsewhere Barth notes that Christians do not have to do with an "earthly fore–court theology or a Christian world–and–life view, but with a Word of God, thus with the approach and way of knowledge epitomized in the confession: I believe." [14] Barth's confessional position displays his aversion to any form of natural theology, and is thus not so much a statement about providence but about pluralism. Having discussed pluralism, I would suggest that Barth has in mind a kind of modern pluralism, one that develops a generic religion out of the religions; but our approach here has been to reject that modality and instead develop something akin to what Barth may have preferred – a Christian doctrine, albeit one that is informed by the world religions and rooted in pneumatology. [15] Secondly it is clear that our approach recognizes the differences between the various religious viewpoints, and in no way are we negating the "specifically Christian doctrine" of providence. Thirdly, we are attempting to flesh out the full implications of general revelation, the rule of God which Barth describes as "the theater and instrument of His righteous acts, the mirror and echo of His living Word, the analogy of the kingdom of heaven." [16]

As we follow our model, the next task before us is adaptation. We have already highlighted some areas that may be of use to us from other religions in the arena of preservation – Buddhism's emphasis on the other in defining our ontology as humans; Hinduism's focus on the immanence of the deity; and Islam's attention to the daily continuous creative work of God. How can we adapt these concepts to a Christian doctrine? [17]

First of all, let us begin with the radically dependent ontology of Buddhism. While we must reject the non–theistic understanding of the origination of humans in crafting a Christian doctrine, nevertheless there is an element in Buddhism here that can be *adapted*, (bringing in ideas more or less wholly) to use our terminology. Certainly in the literature on the doctrine of creation generally, and providence and preservation, specifically, there is scant mention of the role of the other, of the community. How is the other, the community used in the preserving work of God? Here Martin Buber's seminal work *I and Thou* provides some context for us. In contrast to the "I–it" relationship between persons and things, the "I–thou" relationship bespeaks of the person encountering their own being in the 'other': "The concentration and fusion into a whole being can never be accomplished by me, can never be accomplished without me. I require a You to become; becoming I, I say you." [18] The need for the other has become a significant theme in Christian anthropology. Ray Anderson builds on the work of Karl Barth in a chapter entitled "Humanity as Determined by the Other" in his book *On Being Human* (1982): "Thus the word "Thou," although it is a very different word, is immanent to "I". It is not a word which is radically alien, but one which belongs to it. The word "I" with which I think and declare my humanity implies as such humanity with and not without the fellowman." [19] In discussing transcendence as personal encounter, Paul Jewett writes of how the "far–offness" and "nearness" of the other are inseparable:

> My world is a self–contained, unbroken continuum – as is yours. It embraces all that is in the objective world of three–dimensional space from galaxies to quarks – as does yours ... There is, then, no wall of partition where your world ceases and mine begins ...The will [however] is the particular form of the ego 'beyond' the realm of objects; the will is 'where' the I encounters the thou ... So the cognitive–volitional self, the I, stands not only over against the objective world of things but also over against the personal world of the cognitive–volitional thou. Thus I remain forever in my world as you remain in yours. For the self, then, the world of the other is unapproachably 'far–off' and at the same time 'near at hand'. [20]

This sense of dependence, of defining the I through the thou, does not seep its way into the core of many systematic theologies, however, particularly in the area of creation. One theologian who does make use of these themes is Stanley Grenz, whose entire goal in his one–volume systematics, appropriately titled *Theology for the Community of God* (1994), is community:

> In short, the vision of the Scriptures is clear: The final goal of the work of the triune God in salvation history is the establishment of the eschatological community – a redeemed people dwelling in a renewed earth, enjoying reconciliation with their God, fellowship with each other, and harmony with all creation. Consequently, the goal of community lies at the heart of God's actions in history. And God's ultimate intention for creation is the establishment of community. [21]

Grenz further augments his position by noting both the divine image as a shared, corporate reality, fully present only in community, and the roots of the doctrine, found in the creation narratives themselves. [22] As the first human pair is created, the woman is designed to deliver man from his isolation, and as such, the man and woman are utterly dependent on the other for survival. Here we have an example of "dependent creation," rather than "dependent co–origination!"

The Buddhist vision pushes our Christian doctrine to reconsider communal aspects that are already there in our own creation narrative and story, just as the Ancient Near Eastern stories and religions pushed Israel to re–consider/re–emphasize aspects of her own story. This radical sense of dependence, of community, is absolutely essential to a Christian doctrine of preservation, though it has seldom been highlighted. Certainly God's governance, and his work in and through creation is still significant; but the human community (not alone, but through the Holy Spirit) – and its sense of dependence on the other – is vitally important to that understanding of preservation.

If this Buddhist perspective nudges us to reconsider the role of the community in preservation, what of *adapting wholly* the Buddhist notion of dependence into a Christian theology? Can we not suggest unequivocally that without the other we would cease to exist? To do so in totality would be to usurp the role scripture claims for God, who alone holds the universe together (e.g. Ps. 104:24ff: "How many are your works, O Lord! In wisdom you made them all; the earth is full of your creatures … these all look to you to give them their food at the proper time. When you give it to them, they gather it up."). Nevertheless articulating a clear and vitally significant role for the other in preservation is not alien to scripture. The role of the new eschatological community in Christ, described by Paul as a body, is one of utter complementarity: "The eye cannot say to the hand, 'I don't need you!' And the head cannot say to the feet, 'I don't need you!' … If one part suffers, every part suffers with it; if one part is honored, every part rejoices with it" (I Cor. 12:21,26). The Christian doctrine of the body of Christ has typically stopped short of saying what Buddhist doctrine advocates – that we are dependent on the other to exist. Yet scripture supports this position. We have been too cautious in our unwillingness to state things in this manner, perhaps because of our Western individuality, or perhaps because of an under–appreciation of the texts on the role of community in the Bible.

Perhaps there is no finer example of this sense of dependence in recent theology than Miroslav Volf's *Exclusion and Embrace*, which is a practical working out of the preserving work of God in the cross, the self, and the other. Volf outlines his vision: "The practice of 'embrace,' with its concomitant struggle against deception, injustice, and violence, is intelligible only against the backdrop of a powerful, contagious, and destructive evil I call 'exclusion' and is for Christians possible only if, in the name of God's crucified Messiah, we distance ourselves from ourselves and our cultures in order to create a space for the other." [23] "Embracing" the other, and our need for the other must be present in our tomes on creation, lest our systematic treatises unwittingly emphasize the radical individualism and romantic idealism for humanity that was characteristic of modernity. Moreover, by unconsciously relegating the work of God to something transcendent and not something as immanent as the human community, we run the risk of a docetic theology that is subject to all of the criticisms of Feuerbach a century and a half ago. We need *God* to be preserving the human community, and recognizing our radical dependence on each other is a helpful place to start. In this arena we can thus see Buddhism's potentially significant contribution to Christian theology.

A second area from where we can helpfully take our cue is Hinduism's emphasis on immanence in the god Viṣṇu. Here we must note the level of adaptation to be *absorption* – that is, here is an idea already enveloped and developed in Christian theology. We noticed that the presence of an immanent god such as Viṣṇu did not translate into a corresponding sense of community, of dependence on the other. One would think that an immanent god would translate into even more community, or sense of obligation or dependence, but that is not the case in Hindu theology. [24] But nevertheless, the immanence on the part of the deity is there in Hinduism, and acutely so. Immanence has been seen by some as somewhat of a negative category in describing God, one that connotes some kind of reduction of the divine from a rightful place of transcendence. But good systematic theology has always taught that there should be a balance of the transcendent and immanent in the triune God. This balancing point has been truly crucial for most discussions on the nature of God, for to turn too heavily to the side of transcendence invites a distant, uncaring, and even cruel God and universe; turning to heavily to immanence leaves not enough distance between creator and creation, and so shrinks the role of the creator significantly.

One theologian who has consistently emphasized the role of the immanent creator is Moltmann, especially through his work on the role of the Spirit, which we have noted earlier. Moltmann's emphasis on the immanent work of the Spirit is key to his doctrine of creation, and he is concerned to build unity between the creator and creation, and so use immanence to combat the tendency to deism. Another theologian who has articulated a prominent role for the Spirit in creation is Wolfhart Pannenberg. He writes, the Spirit "is the principle of the creative presence of the transcendent God with his creatures: on the other side he is the medium of the participation of the creatures in the divine life, and therefore in life as such." [25] While the Son accounts for the independence of creation and creatures, the Spirit "is the element of the fellowship of the creatures with God and their participation in his life." [26] To explain this Pannenberg later attempts to flesh out this principle via a scientific understanding of "fields of force", which tries to gather simultaneity and space into one concept. [27] While this is a laudable foray into the linkage of science and theology, and a movement away from a static Trinitarian concept into a more dynamic one, we must still see the Holy Spirit's work as more of a personal and creative work, rather than a mechanistic, material one. Still, Pannenberg's inclusion of the Spirit is an improvement over Barth, whose pneumatology in his doctrine of creation is quite limited. Both Pannenberg and Moltmann have carved out a significant role for the immanent Spirit in creation, and this is to be appreciated. [28] We must also mention Moltmann's analysis of the Sabbath here, which we have discussed earlier. Not only does Moltmann explicitly link Christian and Jewish doctrines at this point, his model of divine participation in life is applicable across religious systems.

We must also make reference in regard to immanence and the role of the Spirit to our above discussion of natural theology. If the Spirit is truly moving in all parts of (worldwide) life, then life acquires a sacral dimension to it. Everyone, and even everything, displays the movement of God's Spirit, and particularly in humanity do we see this through the *imago dei*. In a roundabout way, we can come full circle to the need for dialogue with those in other religions simply because of the immanence of God in creation. Of course, this may have been the primal impulse for the entire enterprise of pluralism anyway. But we are still left with the practical import of the immanence of God in day–to–day life, and in fact this is perhaps where the doctrine has the most value. C.S. Lewis' masterful essay on "The Weight of Glory" highlights the sacral life related to the immanence and image of God:

> It is a serious thing to live in a society of possible gods and goddesses, to remember that the dullest and most uninteresting person you can talk to may one day be a creature which, if you saw it now, you would be strongly tempted to worship, or else a horror and a corruption such as you now meet, if at all, only in a nightmare. All day long we are, in some degree, helping each other to one or other of these destinations. It is in the light of these overwhelming possibilities, it is with the awe and the circumspection proper to them, that we should conduct all our dealings with one another, all friendships, all loves, all play, all politics. There are no *ordinary* people. You have never talked to a mere mortal. Nations, cultures, arts, civilizations – these are mortal, and their life is to ours as the life of a gnat. But it is immortals whom we joke with, work with, marry, snub, and exploit – immortal horrors or everlasting splendors. [29]

As we close this sub–section, however, we must ask what is the role of immanence specifically in the doctrine of preservation? Beyond highlighting the role of the Spirit, and the sacral dimension of life, there is another area which deserves mention, and that is the immanence of the divine godhead itself, brought to the fore through Jürgen Moltmann's work. Moltmann's scheme places the "scandal" of the cross within the Triune God itself, and not just in the economic salvific activity of God the Son. The divine godhead's activity shows its essence completely in the history of salvation. And the reason this is so is because of love. For Moltmann, that God is love precludes any possibility of remoteness or distance from the creature, for love embraces, feels, and does not leave any room for impassability:

> The notion of an immanent Trinity in which God is simply by himself, without the love which communicates salvation, brings an arbitrary element into the concept of God which means a break–up of the Christian concept .... It introduces a contradiction into the relationship between the immanent and the economic Trinity: the God who loves the world does not correspond to the God who suffices for himself.[30]

This is how Moltmann can write a book entitled, *The Crucified God* (1974), because for him the distinctions between the immanent and economic Triune God are gone.[31] God is who he is in his actions for and toward us. In particular, this ethos of suffering shows the kind of God Christians worship and is a powerful tool of identification and solidarity. Moltmann concludes, "God and suffering are no longer contradictions; rather God's being is in suffering and the suffering is in God's being itself, because God is love."[32] Thus, "the history of the world is the history of God's suffering."[33]

This message of immanence is one that needs to be shouted, particularly in a time of distrust toward Christianity in the world. This emphasis is precisely what is suggested to us from Hinduism's constitutive immanence, and is one that needs strengthening not simply as the pendulum swing from a long–term accentuation on the transcendence of God, but as a response to what God is revealing of himself in general revelation. An insistence on the immanence of the Christian God is not only encouraged by the perspectives of other religions, it also reflexively gives us a platform for dialogue with other religions. Moreover, it can deeply enhance our own faith, particularly as we follow the priorities laid out before us by Moltmann and others.

The third area to highlight in this area of preservation is Islam's contribution, particularly in the area of seeing each new day, each new act, as a creation of God. This idea can be *assimilated* (modified) into Christian doctrine. Initially, we must return to the issue of "continuing" versus "continued" creation. In choosing the term "continuing" over "continued" creation, our motivation is clear: we wish to underscore the continuing work of God in creation. In rejecting any sort of Deist understanding, our aim is the preservation of the freedom and activity of God, and the consequent rejection of any contingency put upon the creator by the establishment of an on–going ontological necessity. Schleiermacher wrote of the creation of the world as one divine act, and included in it is the whole nexus of nature. [34] "The doctrine of Creation," Schleiermacher notes, "is completely absorbed in the doctrine of Preservation." [35] In Schleiermacher, God is not free to act, and so the preserving of the world includes little activity on God's part. This makes sense given Schleiermacher's attempt to engage the dominant Newtonian worldview of his time – that God acts through the laws of nature.[36] Gunton surmises that "Schleiermacher at once deprives the world of its relative independence and God of his freedom of action within it. He is really only interested in creation as a function of our present experience of absolute dependence..." [37] Schleiermacher is concerned to preserve the unity of God's actions, but in the end cannot avoid reducing not only God and creation, but also the active preservation of God.

What we are interested in is a theology of preservation that stresses God's active engagement with the world. This involves both a linking and a separating of creation and preservation, the latter of which Schleiermacher did not do. By linking, we must ground the preserving work of God in the initial act of creation – "it establishes not merely the temporal beginning of creaturely existence but even this existence itself in all its range." [38] This constitutes a robust doctrine of creation, something needed to establish the Christian nature of creation. But separation of creation and preservation is also necessary, not only in the area of temporality (i.e. beginning and continuance), but to give the creator the possibility of new creations. Here is where the linking of preservation to the original creation becomes important, because in positing the possibility of new creation, we do not want to imply a discontinuity between the original and new creation. [39] All find their unity in the same being and activity of the one God. That God is about new beginnings, even new creations – this aspect of preservation becomes crucial as we take Islam as a springboard. Pannenberg suggests, "the activity of creation bursts forth again out of preservation. Martensen added that we see this bursting forth of creativity in world occurrence wherever we cannot view something as a mere repetition of what has gone before, but that which is new and original manifests itself in its existence. He rightly related to this the concept of miracle." [40]

This present activity of God means that a term such as "continuing creation" is preferable to "continued creation". Though Weber has stated his worries about such a term (see above), and in fact advocates rejecting both terms altogether because of their "obscuring effect," I believe the term captures well the present creative activity of God, which coincidentally, is another needed corrective to a perhaps too–transcendent understanding of God. This creative aspect to God is crucial to exploring the "new" work of God: By grounding the continuing work of God in the *theos* itself, we can both be tethered to the Christian tradition and embark on new journeys in theology. An immanent God is an active God, and an active God is a preserving God. These aspects of preservation will continue to be crucial for any pluralistically–informed Christian theology, particularly if we are attuned to the general revelation present in Islam.

In conclusion, Buddhism's ontology of connectedness should be *adapted*, or brought in wholly, to Christian doctrine as a helpful corrective to an individualistic emphasis. Buber, Grenz, and Volf help us here, but none has outlined the radically communal nature of humanity in the way that Buddhism has, or that scripture suggests. Similarly, Hinduism's suggestion of a strong immanence in preservation should be *absorbed*, or enfolded into, already existing Christian doctrine. Moltmann's immanent thrust in his theology provides perhaps a beginning point, as does his re–highlighting of the role of the Spirit in everyday life. But further probing is needed, particularly in the area of creation. One implication of this emphasis is seen in environmental concerns; if God is intimately involved in all of creation, then stewardship cannot be classified as "issue" theology but must in fact be central to a doctrine of creation. We will have more to say about this later in this chapter. Islam's contribution in seeing each new day or act as a creation of God should be *assimilated*, or modified into Christian doctrine. That is to say, each new act of the "continuing creation" reflects the continual divine working in creation, a sense well articulated in Islamic thought. This emphasis will result in more openness to the new divine workings in creation, and perhaps open us up to new trajectories in theology. And though many theologies are keenly interested in innovation, our position is that grounding such innovation in general revelation positions the theology better for assimilation into the worldwide life of the church.

Transcendence

Thus far in discussing preservation, there has been a fair bit of emphasis on the importance of immanence, which would seem to mitigate against an accentuation of transcendence in our theology of creation. But in fact as we study the world religions, not only Islam, but also Hinduism and Buddhism hold a key place for transcendence in their systems. While that may seem a bit surprising, it shouldn't be, given the balance in this arena of transcendence and immanence that all these religions attempt to hold. We will begin by revisiting some of our findings in the area of transcendence.

Transcendence in Buddhism, as we have already noted, is a curious, yet undeniable occurrence. Nirvana is posited as a completely transcendent ideal for the adherent, an ideal that has at its core, nothingness, but is at the same time the Absolute, and the Ultimate. Nirvana is above all dualities, and even above all description, yet does not include the idea of a divine agent. This lack of a personal being, or "God", should not lead us to see a complete lack of transcendence in Buddhism. In fact, it should open us up to new ways of thinking about God and transcendence – that God is a person, as well as not–a–person, that God is a creator, as well as not–a–creator, for example. This nothingness, or emptiness, at its core will be something we explore as to its usefulness for our Christian theology on creation.

It is clear that in Hinduism there are many gods and many forms and expressions of religious devotion and theology, so that choosing one as exemplary is an impossible task. However, recent studies of Hinduism have begun to re–classify how Hinduism is conceived, from polytheism to *henotheism*. This both underscores our point that transcendence is important to Hinduism, and at the same time, makes our task of learning from Hinduism easier. We do not need to choose one example whose analysis becomes irrelevant when we pick another example; our task is to learn from the One God behind the many *devas* of Hinduism. Here is Klostermaier's analysis:

> R.N. Dandekar has quite convincingly shown that behind the Vedic deva worship lies such an idea of an all–pervading ultimate power, of which the *devas* partake without being quite identical with it... We find it quite acceptable that at different stages different *devas* were at the center of Vedic religion. In our present Veda text

> Indra dominates, a rather late arrival in Vedic religion … Only if Indra is understood as a symbol for, and a manifestation of, the High God do the attributes given to him make sense, especially because the same attributes are given to other *devas*, the same deeds are ascribed to Agni, Soma, Varuṇa, and others.[41]

Klostermaier concludes his section on the gods of Hinduism by noting, "among the *sahasranāmas* – the litanies of thousand names, which are recited in honor of each of the great gods – the overlap is considerable: each one would be named creator, preserver, destroyer of the universe, each one would be called Truth and Grace and Deliverance. Each one, in the end, is the same: One."[42]

If then, for the average Hindu adherent, the transcendent, and the presence of the transcendent, plays such a large role in their theology, how might its notion of transcendence support and/or challenge Christian theology and practice? In particular, how can transcendence be retained in the divine plurality of the Trinity?

Islam's perspective here is the one that is most similar to ancient Israel and Christianity's monotheism. Allah is singular; there is nothing or no one like him and he is to be worshipped alone. But there is also no mediator as in Christianity, and Allah's revelation is seen verbatim only in the Qur'ān. There can be no questioning the heavy emphasis on transcendence in Islam. However, one of the difficulties that arises here in this context of transcendence is monism. How does one avoid making everyone and everything merely an extension of the deity. Is there freedom for the human in Islam's transcendence? Does creation have any autonomy? These are some of the sub–issues that Islam nudges us to explore in turning to our own Christian theology.

Following our methodology outlined above, after we have taken our cue from other religions, we must ask what are the similarities and differences between these other notions of transcendence and the Christian perspective? In Buddhism, we see the lack of a divine agent as well as the unique ontology of nirvana, both of which are completely different from the Christian worldview. However, behind those Buddhist doctrines is the notion of transcendence, which plays a key role in Christian theology. The challenge from Buddhism, as elucidated by Din, is, how can Christianity perhaps incorporate emptiness into its perspective? Hinduism also suggests a strong role for transcendence behind its many gods. Without a polytheistic structure, Christian doctrine outlines the triune God as the manifestation of both transcendence and immanence. What is the relationship, then, between transcendence and immanence? Islam's insistence on a vigorous transcendence for Allah calls into question the role of creation and humanity. Does humanity have freedom or autonomy, or are humans merely extensions of the deity? Here we will need to explore how the freedom of the creature relates to the control and transcendence of the creator. Thus each of these religious traditions poses an important query to Christian theology.

Using our model, with the points of connection with other religions suggested, we must now move to the seminal areas of Christian theology in this area of transcendence. It is important to note that for Christians, the concept of transcendence is at its core a statement about the *relationship* of the creation to the divine. In discussing the nature of the divine, it is quite natural to do so in terms of God's relation to the world, because we have no other vantage point from which to discuss God. [43] So we then have the idea that God is "transcendent" toward creation – that he is related to it as prior to it, both temporally (though not solely in a linear sort of temporality), and ontologically; apart from it (ontologically and actually), and exalted above it (in terms of priority)(cf. Is. 57:15, Ps. 97). [44] This transcendence does not quash God's immanence, which we have discussed earlier, but instead the two are to be held together as explicating two parts of the same God.

How this transcendence has been elaborated within Christian theology has typically been through discussions of the attributes of God. Words such as infinite, eternal, omnipotent, *et al.*, have been used. Pannenberg writes, "In virtue of God's immensity, his presence transcends all that is made. Even heaven and the heaven of heavens cannot contain him (I Kgs. 8:27). Isa. 66:1 finely depicts heaven as not merely the dwelling but the throne of God. God transcends it even though he also touches earth with his feet. Thus God comprehends all things with his presence but is not comprehended by any." [45] Tillich's description of transcendence focuses on the distance involved in transcendence:

> As the power of being, God transcends every being and also the totality of beings – the world. Being–itself is beyond finitude and infinity; otherwise it would be conditioned by something other than itself, and the real power of being would lie beyond both it and that which conditioned it. Being–itself infinitely transcends every finite being. There is no proportion or gradation between the finite and the infinite. There is an absolute break, an infinite "jump." [46]

The practical import of the doctrine of transcendence for Christian theology has been to preserve the freedom and unity of God over against creation.

God is not dependent or contingent in any way upon his creation. From an emphasis on transcendence has followed either a rejection or qualification of monism, pantheism, panentheism, and for some, becoming (i.e. process thought). Once transcendence has been allowed to slip away, the essence of the divine has been seen to slip away as well. And then, according to Jewett, Christianity is left to secularism and ultimately, irrelevance:

> This loss of the transcendent God threatens the entire universe of the theologian (and the universe of everyone else for that matter) with irrelevance. Many theologians have failed to realize the scope of the problem and have assayed to resolve it simply by transposing the traditional statement of the Christian faith into a secular key. Such efforts are doomed to failure, for even though the language of otherworldliness is retained, it is transferred to this–worldly referents ... Thus, in the end [the theologian] finds herself talking not only to secularists but as a secularist.[47]

Now whether the situation is as bleak as Jewett outlines is debatable, but nevertheless the motivation behind preserving the transcendence of God has been important.

All of this "otherness" has led many to wonder if any predication of God is even possible. And here theologians have argued that we can know what God is really like in God's own self–revealing toward creation: God truly is who he is in his revelation towards creation. Barth helps us here in arguing that "God's essence and work are not twofold but one."[48] We can only know the essence of God through his work, exemplified primarily in the witness of scripture to his acts, as creator, reconciler, and redeemer. Thus the work of God *toward us* reveals who he is in essence. This posture – toward us – is seen then in further explication of some of the grand attributes. "We have to see God's immensity and his omnipresence in conjunction," comments Pannenberg. "Precisely as the one who incommensurably transcends his creation, God is still present to even the least of his creatures."[49] Tillich notes that transcendence has a freedom which may conflict with our freedom as creatures, but also has the possibility of reconciling our finite freedom with God's infinite freedom.[50]

There have been some attempts to restate the doctrine of transcendence, in light of some of the problems associated both with the "otherness" of the doctrine, and the demise of concept itself. Bonhoeffer argued for a "this–worldly" transcendence and a "religionless Christianity" to appeal to modern humanity.[51] Kaufman proposes seeing in the encounter of another "thou" an experience of transcendence akin to the divine/human encounter, where the transcendent One makes us aware of himself by disclosing himself to us as the divine "Thou." [52] But rather than recasting the doctrine away from religion, or redressing it towards personal encounter, I would suggest that what is needed is a re–capturing of the magisterial sense of transcendence. And this is held out to us in these other religions. It is significant that Bonhoeffer, and to a lesser extent, Kaufman, are addressing *modern* humanity. Postmodern humanity does not need to shy away from religion, or expressions of religion, or redirect religion as humanism. The narrative and the tradition are present, and should be referred not apologetically to modern humanity, but enthusiastically to postmodern humanity. Transcendence perhaps needs to be re–discovered for our age, and the world religions are pushing us to do just that. [53]

The next step in our model is noting the level of adaptation. We have already outlined some Buddhist, Hindu, and Islamic emphases: Buddhism – the need for emptiness in transcendence; Hinduism – the need for transcendence in a plurality of the divine; Islam – preserving the freedom of the human while still allowing for strong transcendence in the creator. How are these sentiments to be adapted into Christian thinking?

Buddhism's paradoxical thrust cultivates both transcendence, through the attaining of nirvana, and emptiness, through nirvana being pure nothingness. Buddhism is loathe to make attempts to describe nirvana, seeing it as beyond description, emptiness at its core. This understanding can be *absorbed* by already existing Christian doctrines because Christianity knows of the difficulty in describing God as well: "My thoughts are not your thoughts, nor are your ways my ways … As the heavens are higher than the earth, so are my ways higher than your ways and my thoughts than your thoughts" (Is. 55:8–9). And in discussing transcendence, we would do well to emphasize the truly *transcendent* nature of transcendence. Thomas Aquinas saw this in the limitations of understanding God. "The mind is found to be most perfectly in possession of the knowledge of God," he wrote, "when it is recognized that His essence is above everything that the mind is capable in apprehending in this life." [54] This "otherness" of transcendence forces us to recognize that what we know about God is only what he has revealed in turning "toward us". All talk of the nature of God is to be preceded by acknowledgement of God's great distance from the human, and by the bridging of that distance only by the creator coming to us in revelation and mediation, preeminently in Jesus Christ. [55]

But secondly, Buddhism also posits the paradox of emptiness. This too can be *assimilated* (modified) and *absorbed* (enveloped by an already existing doctrine) into a Christian perspective. In suggesting negation as a way to refer to the divine, Din sees process theologians as helpful in relating Christianity to its Buddhist counterparts. Rather than referring to the Christian God as a personal being, Din feels that attempts at describing God as the "Ground of our Being" (cf. Tillich), or the "Process of Becoming" are more accurate ways of describing Christian transcendence, and also better relatable to Buddhist conceptions. [56] Din offers that we should then be open to speaking of God in non–personal terms. The other correlative issue is that it is difficult to speak with absolute certainty as to the nature of God; so since it is difficult to predicate anything and/or personalness to the divine, the categories of negation – i.e. not a person, not a creator, not a thou – should be used.

However, our sense is that negation, in terms of ontology, should not be used as a category to describe the nature of God. There is far too much positive description in the tradition that may encompass the categories Din is referencing; for example, the idea of an "emptying God." The Christian narrative is replete with multi–faceted ways of describing God, so that negation does not need to be resorted to. [57] Robert Jenson's comments echo our own:

> A straightforward doctrine of God's nonbeing must anyway be dangerous. Once the concept of being is there, with any sort of spiritual power, the concept of nonbeing will also be there, with a spiritual power of its own that perhaps cannot easily be manipulated to Christian purposes. Within the actual modern tradition, as it stems from Nietzsche and Heidegger, nonbeing is evoked as mere negativity, that is, as *violence* upon what is; Heidegger's facism was no accident or coincidence. It will, one may think, be best for theology not to allow this dialectic any conceptual terms. [58]

Further, the notion of a personal God cannot be jettisoned. Perhaps it would be clearer to speak of the personal *being* of God, in that it gets to the heart of the matter, that God is a being of three persons, the term persons being more properly defined in the area of Trinitarian studies. Personal, in this day and age, connotes something far too trivial and limited, than what is meant by the term. Tillich understands the difficulties of the term, "personal God", and so puts forth his own understanding:

> "Personal God" does not mean that God is *a* person. It means that God is the ground of everything personal and that he carries within himself the ontological power of personality. He is not a person, but he is not less than personal. It should not be

> forgotten that classical theology employed the term persona for the Trinitarian hypostases but not for God himself. God became "a person" only in the nineteenth century, in connection with the Kantian separation of nature ruled by physical law and from personality ruled by moral law... "Personal God" is a confusing symbol.[59]

While we want to avoid depreciating the concept of God, given our analysis earlier in this paper regarding the God of Israel as compared to the gods of her neighbors, retaining the "personal" as part of the nature of God is essential. It is what separates Yahweh from his counterparts – that he personally *relates* to his people. This sense of *relationship* gets at the core of what it means to be personal in Christian transcendence. Paul Jewett summarizes this sense wonderfully: "The subject of theology is not the God who is the 'Unmoved Mover,' '*Prima Causa*,' 'Ground of Being,' 'Source of Human Ideals,' 'Vital Urge in the Evolutionary Process,' 'Ultimate Integrating Principle,' etc., but the personal, loving God who speaks to us and who says: 'I am the Lord your God' (Ex. 20). The God of Scripture is not a god defined by abstract thought but the God who defines himself as he discloses himself in personal encounter."[60] In fact, God can be love only *if* personal.

Returning to the notion of the emptiness of God, rather than positing a negative of God (i.e. not–a–being), perhaps it is better to note the emptiness already present in the Christian God. And by this we mean the theme of *kenosis*, which Abe has explored in great detail. Abe suggests that only in total self–emptying is God truly God:

> This means that kenosis or emptying is not an *attribute* (however important it may be) of God, but the fundamental *nature* of God. God is God, not because God had the Son of God take a human form and be sacrificed while God remained God, but because God is a suffering God, a self–sacrificial God through total kenosis. The kenotic God who totally empties Godself and totally sacrifices Godself is, in my view, the true god. And it is precisely this kenotic God who thoroughly saves everything, including human beings and nature, through self–sacrificial, abnegating love.[61]

Abe is aware of the trend toward pantheism here, and is also very sympathetic to Moltmann's work in *The Crucified God*. Nevertheless he feels that this needs to be emphasized in finding common ground with Buddhism. We would say in response to this that, first of all, in a purely societal, global sense, this tack regarding the nature of God is quite needed. A too–transcendent view of God has often produced a God of "ruling", of "subjugating", of "self–sufficiency" (and this incidentally, is the view of much of the world regarding Christian religion in general). And while God may rule the world, his rule is not one of subjugation – that comes in reference to evil – but rather of sacrificial love for the world. How is this to be represented in Christian theology? Abe perhaps goes too far (later he suggests the Trinitarian God be viewed as the "great zero"), but an appropriate Christian theological interpretation is seen in Moltmann's response to Abe. Moltmann argues for a social doctrine of the Trinity (cf. *The Trinity and the Kingdom* {1980}), and for *perichoresis*: "It is the divine being of the Father, the Son, and the Holy Spirit to surrender entirely to the other person and in this way to achieve self–realization only in the other persons. The unselfishness in the eternal love and unity of the Trinitarian God is *perichoresis*: community in mutual interdependence and interpenetration." [62] Can we begin, at Buddhism's impulse, to see the social, interdependent, and interpenetrating nature of the Trinity? Moltmann cites the modern Anglican "Hymne to the Creator" as an example of the nature of God seen from this perspective:

> Thou art God, no monarch Thou,
> Throu'd in easy estate to reign;
> Thou are God, whose arms of Love
> Aching, spent, the world enstain. [63]

Can this view of the transcendent Creator begin to permeate our understanding, and so our mission, work, and relationships? We believe it could.

We now turn to Hinduism, and its thrust of transcendence present even in a plurality of gods. This emphasis can be *absorbed* as well into already existing Christian doctrines. Naturally, this turns Christian doctrine toward the area of the Trinity, and while Trinitarian studies may not usually be a focus of a doctrine of creation, still the issues involved relate specifically to what kind of God is involved, and what kind of cosmology is adopted.

In the first half of the 20th century, Karl Barth, concerned to preserve the unity of God – that who God is *ad extra*, is also who He is *ad intra* – posited that God's work shows his essence, though he was careful not to fully identify one with the other. Barth thus takes us down the road of identifying God's essence through his act. The question which encounters us at this junction is, has Barth proven his case that act leads to essence, or more specifically, can act be essence? Another way of restating this question with the tradition of Christian theology behind us is what is the relationship of the economic Trinity to the immanent Trinity? [64] Does one show the other? Is one the other? To answer these questions, we must turn not only to Barth, but to other theologians as well. [65] This Trinitarian analysis will help us to see more precisely the relationship between transcendence and immanence through the threefold persons and activity of the Christian God.

Eberhard Jüngel addresses the "how" of the unity of God that Barth finds so important in Trinity as act. If God's internal relations correspond to His relations with the world, and this unity cannot be broken, Jüngel argues for a unification, rather than an undifferentiated unity. And this unification is seen through the image of "becoming" – God's relationship with the world is one of "becoming" one with the world. [66] This "becoming" is seen as the world is "taken up" into the event of the knowledge of God as "secondary, subsequent subjects." [67] This concept of "becoming" will become important as we analyze the trajectory of Barth's work here, especially in reference to Process thought. But we must note Jüngel's insistence on humanity as "subsequent subjects". Jüngel is concerned to maintain the transcendence of God, a position put in jeopardy with the blanket identification of act with essence.

Karl Rahner takes Barth and Jüngel further, and expresses singularly and strongly, that the economic Trinity *is* the immanent Trinity. While Jüngel desires to speak of "correspondence", Rahner takes us to unity. [68] Rahner states unequivocally that the economic Trinity is the immanent Trinity and the immanent Trinity is the economic Trinity. [69] Rahner's special emphasis is on each of the divine persons communicating himself to humanity, "each in his own special and different way of personal being, in the free gift of grace."[70] This singularity of communication, as opposed to God in general, is "to be understood not merely as the communication of the divine 'nature' but also and indeed primarily as communication of the 'persons', since it takes place in a free spiritual personal act and so from person to person." [71] So that for Rahner the "personal" is indispensably linked to Trinity as Act, but he would have us focus in on the way the one God self–communicates in three self–communications:

> Hence the Father gives himself to us as *Father*, that is, in and by the very fact that being essentially himself he expresses himself and thus imparts the Son as his own personal self–disclosure: and also in and by the very fact that the Father, and the Son who receives all from the Father, affirming themselves in love, inclining to themselves, coming to themselves, impart themselves in loving acceptation, that is, as Holy Spirit. God's relationship to us is three–fold. [72]

Thus for Rahner, God's immanence is primary. He is for his creatures. The ancient problem of the "otherness" of God relating to the subject of finite humanity is overcome in this self–communication. Rahner does not necessarily need Barth's and Jüngel's unity of the divine *in extra* and *ad extra*, for he concentrates on this self–communication as "the free gift of grace." The divine subject is free to act in whatever way he deems necessary. A major part of Rahner's focus is on the human experience of the Trinity, which he wants to trust as showing us the divine reality. [73] Rahner seems concerned to make the Trinity a part of everyday life for the church, and as such his talk about the Trinity mirrors Barth's emphasis on the work of the Holy Spirit – in the believer. It is almost as if here we have moved from Trinity as Persons, through Trinity as Act, and on to Trinity as Relationship.

But God is still other for Rahner. He is careful to distinguish God's self–communication from anything that humanity can completely grasp, and in so doing, maintains a difference between God's essence and God's work. Rahner states, "It is a self–communication in which that which is imparted remains the sovereign and incomprehensible, and which even as something received continues to be unoriginated and not at the disposal or within the grasp of anyone." [74]

In Barth, God's work reveals his essence; in Jüngel, God's dealings with humanity correspond to his divine interrelations, and God is *becoming* united with the world; and in Rahner, the self–communicating God is *in* us, though not in totality or completely. Rahner has moved beyond both Barth and Jüngel. All three emphasize God's relation with the creature, and all three maintain the gap between the self–moved, transcendent God, who remains the I, and the secondary human Thou. In all three systems, we see the divine God defined in action, in activity, in relationship. The demarcation between act and essence, ontology and economy, is faint, though it is nevertheless there.

The greatest progression of Trinitarian thought, hinted at in Jüngel's analysis, is to be found in process thought, though most process theologians are not concerned with traditional distinctions regarding the Trinity. Lewis Ford is one who is, however, and he suggests that there is a triunity of principles within God, that can be roughly related to the Father, Son, and Holy Spirit of traditional orthodoxy: The non–temporal creative act of God, its outcome within his primordial nature, and the divine experience of the world as his consequent nature. [75] This has similarities with Moltmann's scheme, and shows the complete replacement of the immanent Trinity with the economic one. Moreover, in this economy, the "becomingness" of God has no essence apart from his act, and there is no ontological ground for the event of history, which is so important for Barth. The same criticisms of Hegel may apply to process theology as well in this context. However laudable the attempt at overcoming the transcendent "barrier" between God and humanity, it seems that process theology is doomed in its attempt to relate the basic dual nature of Whiteheadian epistemology to the triparte nature of Biblical Trinitarian thought.[76]

In the end, this Trinitarian analysis has shown that transcendence must remain a constitutive part of Christian theology. Transcendence helps define the divine, its relation to creation, and humanity's relationship to both the rest of creation and the divine. To let go of transcendence would be to lose some of the crucial grounding of Christian ontology. Here Christianity can listen to its own voices and history, while still hearing an affirmative echo in Hinduism's impetus. We must retain transcendence as a significant mooring for a Christian discussion on the nature of God.

Finally, Islam has material for Christian thinking on transcendence to consider. Islam's strong sense of transcendence has been noted, and we have highlighted a number of aspects to the divine transcendence brought to us through Buddhism and Hinduism – emptiness, maintaining immanence with transcendence, etc. One area where Islam struggles, given its ontology of omnipotence, is in the freedom of the human. How can men and women be free to choose their actions and activity if Allah is everything? For instance, A. Guillaume writes,

> There are texts which clearly assert that man is responsible for his own actions, though the majority of texts seem to assert that they are definitely decreed. The Mutazila dealt with these passages as best as they could by softening the language of predestination, but still it could not be denied that the orthodox party had the Quran on their side when they asserted that God's predestination was absolute. This view is borne out by the chapter on predestination in the books of canonical tradition which do not contain a single saying of Muhammad's which leaves freedom of action to man. Everything is predestined from the first and a man's fate is fixed before he is born....Orthodox reaction to the doctrine of free will took rather a strange form. The Mutazilites were dubbed dualists because it was said that by their assertion that man has 'power' over his actions they made him the 'creator' of his works and thus encroached on the almighty power of God, for there would be two creators of actions.[77]

This is an area of transcendence where Christianity does not need necessarily to *adapt* to Islam, but rather can provide a better developed theology – the ontological freedom of creation juxtaposed alongside of the freedom and power of the creator. Here we can note that dialogue can involve listening and learning, as well as speaking and offering. We will approach this topic from two angles; the first is via the creator, the second is through the creature.

In terms of the creator's perspective, it is clear that Christian theology desires, alongside of the creator's transcendence, omnipotence, and omniscience, to preserve the freedom of the creature. One attribute that is not predicated of God is that of omnicausality. Simply because God has all power and knowledge does not mean he causes all things to pass: There must be room for the creature, and for sin. [78] One example of this is the Spanish Jesuit Luis de Molina's construction of "middle knowledge" (*Concordia liberi arbitrii cum gratiae donis* {1588}). In this knowledge (*scientia media*), God's influence on the free actions of the creature is limited to giving them a general bias toward the good and attempting to move them in this direction by commands, counsels, warnings, etc. [79] Thus God is not involved in the free actions of his creatures directly; however, he knows what a person *can* do, *would* do, if certain circumstances were available, and what she *will* do under all conceivable circumstances. The actual accomplishing of the act depends on the union of creaturely will and prevenient grace. The result of this cooperation of will and grace is the conditionally future (*futuribile*), which is the object of God's "middle" knowledge. The countering Thomist position stated that it is impossible that divine foreknowledge of the creature would include something free and independent from the divine will. The creature's actions are included in the divine decree. The question then arises, what of the freedom of the human? The Thomist reply is that "the freedom of the human will is not taken away by God deciding and moving it. On the contrary, it is in virtue of this that it attains to its free action as a creature." [80] Only in the movement of the created will by God does humanity act in concert with its nature: "God, therefore, is the first cause, Who moves causes both natural and voluntary. And just as by moving natural causes He does not prevent their acts being natural, so by moving voluntary causes He does not deprive their actions of being voluntary: but rather is He the cause of this very thing in them; for He operates in each thing according to its own nature." [81] Thus we have two developed ways of conceiving how God in his omnipotence allows for freedom in his creation. [82]

Contrast this subtlety with Cragg's analysis of freedom within Islam:

> The Muslim concept of toleration has been, from the beginning, that of freedom to remain what you were born or freedom to become a Muslim. It has never yet meant freedom of movement of conscience, or freedom to become. It was Muslims, rather than non–Muslims, who had the lesser range of intellectual and spiritual possibility. Islam has been traditionally proud of its toleration based on the famous Quranic saying: "There is no compulsion in religion." Modern Muslim commentators fully share that pride. Expositions of Muslim toleration can readily be found in a score of publications and journals. But the fundamental question as to whether there is personal freedom of the Muslim vis–á–vis Islam is not faced and is often not even suspected as present. It is assumed that Islam is a faith that no Muslim would ever conceivably wish to question. Consequently the option to do so is neither valid nor feasible. It is nonexistent… A true understanding of freedom, as freedom of movement of mind, demands that the option should exist. [83]

Regarding this issue of human freedom and divine transcendence, we then move to the human point of view, and the theological controversies surrounding the role of the human will. Luther, in his *The Bondage of the Will* (1525), likens the human will to a beast of burden, ridden to wherever the rider (God or Satan) wants. [84] Erasmus counters by arguing that it would be ridiculous to implore one towards good or evil if one had no say in the matter. [85] Erasmus thus defines free will: "By freedom of the will we understand in this connection the power of the human will whereby man can apply to or turn away from that which leads unto eternal salvation." [86] Human will needs to be united with God's grace, according to Erasmus: "Now, if man could do nothing, there would be no room for merit and guilt; consequently also none for punishment and reward. If on the other hand man were to do all, there would no room for grace, which is very often mentioned and emphasized by Paul." [87] Luther's strong reply argues for the foreknowledge and omnipotence of God being necessary and controlling. Yet even Luther leaves room for some creaturely activity: "Yet, God does not work in us without us, because He created and preserves us for the very purpose that He might work in us and we might cooperate with Him, whether that occurs outside His kingdom and under His general omnipotence, or within His kingdom and under the special power of His Spirit." [88] Similar spirited debates have been going on for centuries regarding human freedom and divine action. Despite differences, basic to both sides in this debate has been the desire "to ascribe to the creaturely will some kind of autonomy, be it quite small and relative, and at the same time to avoid what appears to be the unavoidable idea that God, if he does precondition the human will in such an all–encompassing manner, is then made into the source of evil or of sin." [89] It would seem, then, that an important facet in ascribing an important place for transcendence in creation would be that of the freedom and autonomy of the created subject. This is an area where Christian theology has a rich tradition to draw upon, one that gives dignity to humanity in their choices without making it robot–like in the face of a transcendent, omnipotent God.

So, by way of summary, in this area of transcendence for a pluralistic Christian theology, Buddhism suggests we build upon the difficulty in predicating anything of the divine. This doctrine can be *absorbed* into already existing Christian doctrines. Buddhism also posits the emptiness of God as crucial. This doctrine can be *assimilated* (modified) into Christian doctrine, and we have seen attempts from Din, Abe, and Moltmann. However, much more remains to be done in this area. Developing a doctrine of "transcendent emptiness" through kenosis should play a key role for pluralistic Christian theology in the area of creation. Hinduism calls for transcendence to be maintained in a plurality of gods. This can be noted and *absorbed* into Christian thought. Naturally, for Christian doctrine, the Trinity has stated the divine plurality and unity, so we explored how the transcendent and immanent can be maintained together in Trinitarian thought. Christianity can also be reminded from Hinduism here to emphasize transcendence while still maintaining the immanence of the creator. Islam's high stress on the transcendence of God pointed us, by way of a limited example, to a necessary corollary that must be maintained – the freedom of the creature. Here Christianity has a more developed understanding that is to be retained – advocating for the freedom of the creature is crucial against the backdrop of the transcendence of God.

Stewardship of the Cosmos

Our final area to consider is that of the stewardship of the cosmos, and by this we mean the human stewardship of the cosmos, as opposed to the divine stewardship – preservation – that we have already discussed. Each religious tradition has a number of ways it advocates for the stewardship of the cosmos. Here we are not simply referencing environmental concerns, but rather the whole of how humanity approaches keeping life going in this universe. Following our model, we will begin with other traditions' contributions to this area, and then see what kind of responses can be made from Christian theology in terms of incorporating those contributions.

As we noted earlier in chapter six, Buddhism's strong ethical concern is to cultivate "skillfull means" as a way to attain moral excellence, and so the main moral focus is on the here and now. An example of this is the Eightfold Path – right view, right aspiration, right speech, right action, right livelihood, right effort, right mindfulness, and right meditation. [90] Moving toward stewardship issues, this ethical emphasis assumes the interconnectedness of humans and all the world: "[Buddhism's] ethic of non–injury and boundless [involves] loving–kindness for all beings. The well–known Five Precepts *(panca sila)* forms the minimum code of ethics to which every lay Buddhist should adhere. Its first precept involves abstention from injury to life. Buddhism also prescribes the practice of *metta*, "loving–kindness" towards all creatures in all quarters without restriction." [91] Certainly through the doctrine of reincarnation, Buddhists are taught to revere all creation and animal life. By living at peace with and in appreciation of nature, Buddhism's environmentalism is grounded in the tradition as well. Brown notes, "Nature's beauty is appreciated by Buddhists because it is recognized as valued by those of high spiritual attainment. The Buddha and his disciples regarded natural beauty as a source of great joy and aesthetic satisfaction. In various ways the Buddha has stressed the need for close contact with nature and pointed out how advancement of mind leads to a greater appreciation of nature. By default one could say that the misuse of nature tends to indicate low spiritual attainment." [92] Another attendant issue here is the no–self of Buddhism – all thought of ego, me, mine, self–pride, is to be jettisoned in the pursuit of the noble path. Just how the lack of a self corresponds to Christian ethics, especially in the context of the *imago dei*, will be something that Christian theology needs to explore.

Hinduism's contribution in this area of the stewardship of creation is its emphasis on the sanctity of life, all life. This is seen primarily through the doctrine of reincarnation. There is a much greater continuity between humans, animals, and the natural creation because of the cycle of rebirth. Klostermaier describes it thus: "The gross body is destined to disintegrate at the time of death. The subtle body with all the imprints of deeds and thoughts of the previous life, is preserved and clings to its *ātman*; this results in intermediate existences in heavens or hells, depending on the quality of the deeds, and finally in an earthly rebirth. Karma keeps in motion the vicious circle of action, desire, reward, and new action." [93] The Christian concept of the *imago dei* again comes into play here and is differentiated from the Hindu conception, which sees no intrinsic divergence between humanity and animals. Our goal here will be to see how the *imago dei* stands in the context of all of creation, and whether such a differentiation (such as between humanity and animals, or between humanity and the rest of creation) is warranted. There may be much in Hinduism that Christian theology can learn from here.

Islam suggests that humanity is God's "viceroy on earth", stewarding the earth; this is similar to the Christian idea of humanity having "dominion" over the earth. While arguing for the important role of humanity in preserving the earth, and acknowledging theological resources in the Qu'rān for the Muslim world, Sardar argues

> The fact that these values, in their environmental aspects, are nowhere to be seen in the Muslim world, where the environmental situation is just as acute as in the West, can be attributed to a very simple fact. Nowhere in Muslim societies is the Shariah [Islamic law] adhered to in its totality or the Islamic way of life forms the basis of human action. Moreover, Islamic concepts and ethical precepts have been divorced from a pragmatic, living, dynamic form for the last 300 or 400 years, since the decline of the Muslim civilization and its eventual colonization by the Occident. It is because the dominant patterns of behaviour, development and thought in the Muslim world are Western that we see this environmental degradation and exploitative way of life. Muslim societies themselves need to appreciate the ecological principles of Islam and find practical routes for adopting and establishing them.[94]

Whether or not Islam's limited environmental concerns are tied completely to its linkage to the West is open for debate. Nasr suggests that Islam is amenable to environmental concerns:

> The ecological and environmental disasters to be seen in many parts of the Islamic world today, especially in the big cities, must not be seen as being the result of Islamic teachings any more than can the terrible pollution in many parts of Japan be considered as the consequence of Zen Buddhism or Shintō teachings about nature. The traditional Islamic view of the cosmos and the natural order did not cause such disasters but created for more than a millennium a civilization that lived in peace with nature. Islam also developed a vast metaphysics and theology of nature and also a traditional art in which nature as God's creation and reflection of His wisdom and power plays a central role without the art ever becoming naturalistic.[95]

Here we have a situation similar to the Christian one in that we have a monotheistic religion, with a transcendent God and an ontology recognizing the reality and significance of creation, struggling to care for the earth. Admitting that environmental concerns are a recent development in theological studies, we are still left with the question of humanity's "dominion" over the earth. How should humanity "rule" the earth? And are there lessons we can learn from the Islamic worldview that will inform our perspective in Christian theology?

Following our model, our next task is to ascertain levels of adaption. Setting aside for now Buddhism's obvious environmental concerns, it is clear that an interpenetrating ontology suggests a strong *ethical* stewardship of creation. Too often in discussing stewardship of the earth, the topic quickly veers toward solely "natural" concerns, in the sense of nature (i.e. trees, forests, water, ozone, etc.). Buddhism reminds us of the role humanity plays in preserving the earth through its ethical actions toward other humanity; and further, Buddhism's strong inter–connected ontology pushes us to adopt a similarly communal Christian ethic. Ethics, for Christian theology, has tended toward "individual" ethics, and not "communal" ethics. Without the context of community, the adoption of individual ethics has lent itself to co–opting by materialist and capitalist concerns. Donald Worster thus argues that the roots of our contemporary environmental crisis go back not to religion, but rather to the individualism of materialism, perhaps best exemplified in the teaching of Adam Smith. Smith argued for a system of "natural liberty", where every person can pursue his or her own interests. The natural world has no intrinsic value to humans, but is rather valuable only when it serves some direct human use or can be exchanged for something else that has value. [96]

However, by viewing ourselves as interconnected, "created for community", as Stanley Grenz puts it, our ethics can have a greater concern than simply for personal wealth. Ray Anderson, following Barth and Buber, outlines a place for "humanity as determined by the other". He writes, 'because the life of the person is structured in co–humanity and is conditioned and determined by the other, there can be no authentic subjectivity without participation in the outward, historical life of co–humanity." [97] This ontology of the "other", in defining ourselves and our humanity, lends itself to a more communal ethic. Grenz grounds this communal ethic in the divine image: "The creation of humankind in the divine image, therefore, can mean nothing less than that humans express the relational dynamic of the God whose representation we are called to be. Consequently, each person can be related to the image of God only within the context of life in community with others." [98] Grenz also finds eschatology important to communal ethics:

> The New Testament envisions a present foretaste of the eschatological human community. The focus of this present experience, according to the New Testament writers, is the community of Christ. As the eschatological community, the fellowship of those who seek to reflect in the present the future reality of the reign of God, the church of Jesus Christ is the prolepsis, the historical foretaste and sign, of the image of God... the fullest expression of our participation in the image of God must await the eschatological transformation of human life in the Kingdom of God. [99]

Thus, almost by way of introduction to the idea of stewardship, and spurred by Buddhism, we must begin by establishing a communal ethic and ontology for humanity.

There is also linkage in this notion of communal ethics to the Buddhist idea of no–self. By subsuming the self within the community, Christianity is attempting to mitigate against the pride and conceit that can attend individualism. However, here Christian anthropology retains the self, which seems preferable to losing it altogether. Anderson asks, "Does the form of co–humanity safeguard against an unwarranted individualism by depriving my own being in its particular singularity of any meaning for me? ... By using the concept of singularity, we intend to preserve the notion of discrete individuality in this structure of determination." [100] The key then is to somehow preserve individual singularity with communal ethics.

So by way of summary, we see correlatives to the Buddhist emphases already present in Christian theology – namely, the turn toward community both in anthropological ontology and in ethics. But we must emphasize the crucial role of ethical stewardship through an interconnected ontology of humanity. Buddhism offers us this perspective. This Buddhist emphasis needs in turn to be *adapted*, or brought more fully, into a Christian understanding. Nevertheless, Christian thought retains, in contrast to Buddhism, the idea of the self as responsible and discrete within the overall community.

Hinduism forces us to re–examine humankind's place in the entire creation. Here it is apparent that Hinduism's doctrine of rebirth minimizes the ontological difference between the human and animal creation, and so causes us to ask what is the role of humanity in creation *vís a vís* this issue of stewardship? Hinduism has, in effect, been on the leading edge of this question for centuries, and consequently, we must see Christian theology as *adapting,* or bringing in ideas more or less wholly, from Hindu thought. Of course Christianity has begun to consider these questions on its own in the last few decades, and so the question is rightly posed, is this not an instance of Christian theology correcting itself? Why need we go outside the Christian tradition to be spurred on by the spiritual wisdom of other religions, by what Christians believe to be a part of general revelation? Part of the answer lies in the temporal realm: had we been paying attention to Hindu thought in this area, perhaps our own theology in this area of stewardship would have developed sooner than in the last few decades. Our theology acquires a sense of *urgency* in dialogue with Hinduism on this point; yes, we have developed our own resources, but Hindu ontology quickens us to revisit our own scriptures.

One place to begin is by asking what is meant by humanity created *imago dei*? This is a question which has been pursued relentlessly in the history of the church, from Augustine, through Aquinas, the reformers, to today. Certainly humanity in its nascent inception begins well, with the divine inbreathing as a distinctive, powerful, and unique part of the creation narrative; and, in fact, the Old Testament speaks of humans as "fearfully and wonderfully made" (Ps. 139:14). Following Anderson's analysis, the church began interpreting the *imago dei* in one of two directions. The first is Thomas Aquinas' doctrine, imbued with a strong Aristotelianism: "Thomistic theology divided the imago dei into three stages: a natural endowment enabling knowledge of and love for God; actual conformity to the image in knowing and loving God; and complete perfection in the love and knowledge of God. 'The first state of the image then is found in all men, the second only in the just, and the third only in the blessed' (*Summa*, I, xciii, 5)." [101] By locating the *imago* centrally in reason, Aquinas is in direct contrast to the second direction, which is a more Augustinian conception. This view locates the substance of the *imago dei* primarily in terms of capacity for relation to God (cf. Luther, Calvin). Once defined, the central issue for the *imago dei* becomes how humanity can retain it in the face of overwhelming sinfulness. So too, then, has theology pursued the relationship of the *imago* to the biblical and empirical fact of humanity being a sinner.

We recount this background data on the *imago dei* to say that though humanity starts from an exalted place in reference to the rest of creation, we must be clear about where it ends up – in need of redemption by God. This starts our thinking about where humanity lies in relation to the rest of creation. Secondly we must note the Noahic covenant (cf. Gen. 6–9). [102] After the flood, a divine covenant is established, and six times the text speaks of the relationship between not only God and Noah and humanity, but also of "every living creature" (cf. Gen. 9:9–17). Bernhard Anderson writes, "The

Noahic covenant, then, is universal in the widest sense imaginable. It is fundamentally an ecological covenant that includes not only human beings everywhere but all animals – every living being (*nepeš ḥayyă*) of all flesh that is upon the earth (9:16 repeating what was said in 6:19)." [103] Obviously, there is warrant for seeing humanity (though it is *imago dei* and the rest of creation is not) as not something *completely* dissimilar from the rest of creation. In fact, it is in need of, along with the rest of creation, God's blessing. Bouma–Prediger notes that the creation of humans does not occur on a different day from the creation of the other animals. There is no separate day for humans, and in fact they are made from the dust of the earth, as earth–creatures. [104]

Thirdly, we must take into account the epistemological context in which we move, and note that much placement of humanity over the rest of creation has its roots in modernity. According to Granberg–Michaelson, "Modern cultural and theological assumptions have placed humanity at the center of purpose and meaning in the universe." [105] Certainly humanity has at times adopted a modern, almost romantic view of autonomous humanity reaching and striving to master life and nature. In some ways, humanity has adopted Nietzsche's nineteenth century romantic vision of heroic man. But we must contrast this viewpoint with the biblical one. While humanity is singularly determined to be the *imago dei*, it nonetheless must deal with the effects of sin, evil, and fallenness, as well as a biblical perspective which places the center of the universe squarely in the *theos* and not the *anthropos*. This is the thrust of Barthian theology as well as the message of Job. In an incisive analysis of the response of God in the whirlwind to Job, Bouma–Prediger notes that the passage suggests: 1) humans are not the center of things; 2) humans are not the measure or measurer of all things; 3) The world is one in which all creatures have a place – even the wild and chaotic, and the dangerous and scary; 4) The natural world displays order and patterned regularity, though it may appear chaotic and unfathomable. [106] This theology of Job tells us that God stands at the center of things. A correct Christian ontological cosmology will place God at the center, as the self–grounding subject, with humanity and the rest of creation as the object of his love. Now it is clear, given the inherent nature of the *imago dei*, that humanity can never be singularly aligned with animals and the rest of nature; there must remain a difference between the those made in God's image and that which is not (though it is all called "good"). And in fact, humans are given the responsibility of naming the other creatures, an important task given the significance of names in the Bible.[107] But nevertheless, far from demarcating sharply between humanity and the rest of creation, we must move in the direction of placing humanity within its rightful context with all of creation as the recipient of God's grace and blessedness. Here Hindu thought can be a great help to us, despite its insistence on reincarnation.

Islam's contribution here is seen in highlighting humanity's role as "vice–regents", or in the Christian parlance, "exercising dominion" over creation. Islam has not developed this role of "vice–regent" in the direction of stewardship of creation as fully as Christian theology has done in the last few decades. Nevertheless, Christian history is replete with examples of exercising dominion in terms of domination and even exploitation of creation, not stewardship. So though Christian thought may be leading Islam on this point now, its recent past is not too far from memory, and contemporary Christian scholarship on this issue must be allowed to sink into the larger Christian ethos. At this point, we must ask, what kind of dominion is envisioned scripturally? It is clear that many view the command in Gen. 1:28 to have dominion and subdue the earth as the cause for a great many of our environmental ills. Arnold Toynbee, for example, argues that the command to have dominion in Gen. 1:28 has permitted and directed humanity to dominate and exploit creation.[108] The solution to this perspective, according to Toynbee, is to move from the "*Weltanschauung* of monotheism to the *Weltanschauung* of pantheism."[109] Roderick Nash suggests that given the harsh wording implied in "domination," "it followed that the Christian tradition could understand Genesis 1:28 as a divine commandment to conquer every part of nature and make it humankind's slave."[110] And of course, Lynn White, in an influential 1967 essay, "The Historical Roots of Our Ecological Crisis", went even further and posited that Christianity itself bears a "huge burden of guilt" for the present (disastrous) ecological situation.[111]

While it is not within our scope to answer White's overall charge, we must nevertheless reflect on what is meant by having dominion.[112] Hinduism's seamless acceptance of all life as sacred forces Christianity to confront its own theology. Dominion does not mean domination, according to Bouma–Prediger, who suggests we look at a larger canonical perspective for what it means to exercise dominion:

> For example, Psalm 72 speaks most clearly of the ideal king – of one who rules and exercises dominion properly. The psalm unequivocally states that such a ruler executes justice for the oppressed, delivers the needy, helps the poor, and embodies

> righteousness in all he does. In short, the proper exercise of dominion yields shalom – the flourishing of all creation. This is a far cry from dominion as domination. [113]

Further, later in Genesis 2:5, for instance, humans are called to "serve" (*'ābad*) the earth, so that any understanding of the task of humans concerning the earth cannot view it only in terms of domination and subjugation, but rather must take into account the entire scriptural picture. [114] Wendell Berry sums up such a limited reading of the scriptural mandate in Gen. 1:28:

> Such a reading of Genesis 1:28 is contradicted by virtually all the rest of the Bible, as many people by now have pointed out. The ecological teaching of the Bible is simply inescapable: God made the world because He wanted it made. He thinks the world is good, and He loves it. It is His world; He has never relinquished title to it. And He has never revoked the conditions, bearing on His gift to us of the use of it, that oblige us to take excellent care of it. If God loves the world, then how might any person of faith be excused for not loving it or justified in destroying it." [115]

Berry's analysis points us toward not only a new understanding of the role of humanity in terms of caring for creation, but also toward an understanding of humanity's role as "vice–regent". The world is not for humanity to destroy, or rule over, or subjugate; the world is not "humankind's slave". The world is for humanity to exercise dominion over *as God's vice–regent.* An improper interpretation of humanity's role in Gen. 1:28 assumes it is free to do what it wants with the earth. This is not so. Humanity is given dominion over the earth *as God's representative*. We must go back to the concept of the *imago dei* for a moment. It is no accident that the call to dominion comes immediately after the description of God's image. Weber cites Zimmerli in describing it as a "call to feudal lordship", and goes on to note, "it is generally characteristic of the Old Testament that it does not make statements about 'nature' and 'being' but statements about 'the task' or a 'relationship'." [116] And so we must view the "task" in mind here – dominion – as related to the image of God in humanity. Humanity is acting in the image of God when it exercises "dominion over" (perhaps better would be – "caring administration of") creation acting in God's stead as his representative. The central issue here is that dominion must be located in the divine ruling of creation, not the human ruling. [117] Von Rad understands it thus: "Just as earthly monarchs set up statues of themselves as symbols of their claim to authority, in provinces of their realms in which they are not constantly physically present, man too in his image is set upon earth as a sign of the dominion of God. He is in act mandated by God, ordered by God to preserve and impose God's claim of sovereignty upon earth."[118]

What we see, then, is a development of doctrine as we dialogue with other religions: A move away from a human centered cosmology to an "all of creation" centered ontology (suggested by Hinduism), but ending ultimately in a divine centered cosmology (suggested by a re–examination of Christian doctrine). In this divine centered cosmology, we encounter a fresh solution to issues of stewardship of creation. God loves the world – he created it, redeemed it, and desires to bring it to fullness as the "new creation". Any Christian doctrine of creation must account for the divine–centered creation, one of love and concern for the world. And here is where we can re–introduce the concept of Sabbath, which we have discussed earlier. The Sabbath shows the divine love for creation at interplay with all of its parts, and Smart gives us a vivid picture of this vibrancy of creation:

> The cosmos is fun. It is, as we have noted, a dance, and the colors and beautiful music and food and water and wine of the world, and the roses and the hyacinths, even the dandelions, are there to be loved and enjoyed. The advent of consciousness spreads out before us, and before God who shares in our consciousness, the vast pleasures of color and form and scents and sounds and all the glories of this world. God saw that it was good, says Genesis; and the dark sense that death is in store for Christ could not blight his enjoyment of the wonders of this beautiful and stormy cosmos… There are the joys of sex, and of eating, and washing the back of the throat with beer and running in the evening air. There are the joys of babies with early smiles, and the whispers of lovers, and tranquil old age, and sports and all sorts of glories.[119]

Far from diminishing environmental concerns, locating the ontology of creation back in the divine will push us to see the true connected nature of all creation. The Sabbath exemplifies for us the "feast of creation", in which the divine, human, and all of creation can come and enjoy the rest of the creator in the world he has made, which is "very good". From correct ontology flows correct cosmology and in this case, the divine love for creation is the wellspring from which flows a correct view of stewardship of the cosmos.

In summary, then, in this area of stewardship, we have discovered that Buddhism's emphasis on ethical stewardship brought about by the interconnectedness of humanity needs to be *adapted*, or brought into the Christian story. Grenz has pointed us in the right direction in this by positing a communal ethics, but Christian theology needs to go even further, and integrate those communal ethics with stewardship. Buddhism can help us here as well through its seamless integration of collective humanity with the natural order. Hinduism has been out in front of Christianity for a long time in the area of seeing humanity as interconnected with the rest of creation (i.e. animals, plants, etc.). Again, Christian thought needs to *adapt*, or bring into Christian theology, this perspective, and it has been doing so in the last few decades. Finally, with regard to stewardship, Christian theology has recently begun to see the role of dominion in terms of serving the earth, and not exploiting it solely for humanity's sake. Islam does not have as much to teach Christianity here. However, by re–asserting humanity's role as *God's* representative (an understanding significant in Islam), we can begin to see what caretaking the earth would look like in the light of God's love for the entire world.

Conclusion

We have now come to the end of our journey, and, just as no trip is complete without the recitations of its highlights, we must also recount our discoveries and conclusions. First, our pluralistic context forces us to ask somewhat different questions in our theologies than did previous generations of theologians. While not necessarily being encouraged to explore the core of special revelation, we are being challenged to investigate in much more detail the *revelation* in "general revelation." The presence of differing faiths next door, down the street, in our classrooms, in the media, precludes us from ignoring their claims. As citizens we cannot ignore the world around us – the days of insularity are over – and neither as theologians can we ignore the world around us. This does not mean that we can merely catalogue more thoroughly the belief systems of others. We must also be open to learn from these belief systems. The reason this is so is because of our second conclusion: both epistemologically and theologically we are driven to do so. Epistemologically, the arrival of postmodernity has coincided with this burgeoning sense of diversity in the world. Postmodernity suggests that rather than a foundationalist sense of knowledge, we need a non or post–foundationalist theory, a web, in which to incorporate our beliefs. The dubious cornerstones of reason and history need to be replaced with a web of belief that considers a wide variety of sources with a core at its center rather than a foundation at its bottom. Theologically, we are driven to consider sources outside of Christian theology because of a tenet of Christian theology itself – natural theology. We have argued that a rightly conceived natural theology – one in concert with special revelation, held together by the work of the Holy Spirit – needs to be re–articulated as our rationale for moving forward in this endeavor. This precludes the jettisoning of natural theology completely (a la Barth), and suggests instead a return to what orthodoxy has always considered a complement to "revealed" theology.

Colin Gunton and Ninian Smart provided two differing doctrines of creation to spur our thinking, even while each was found problematic. Gunton's theology was limited to Christian sources, while Smart's was not. We noted the implications of each approach: Gunton's had little to offer in the way of dialogue, while Smart's fell prey to the difficulty associated with a purely pluralistic endeavor – it didn't do justice to the specificity found in the Christian (or any) religion. A possible improvement to Gunton's schema would be to emphasize further the work of the Holy Spirit in his Trinitarian analysis. This might open his theology to other non–biblical sources, as well as diffuse the sense of Christomonism present in his presentation. Smart's system could be improved by maintaining more of an adherence to a (Christian) tradition. Given the problems associated with either approach, we argued that a third way must be developed – one that is self–consciously Christian, rather than a pluralism, yet one that incorporates material from other religions as well. This would be a truly pluralistic *Christian* theology, one capable of addressing today's context with the truth of God's full revelation to us.

Being interested in using material from other religions, we revisited historical theology, to see if there were any resources for us in learning how to do that. Surprisingly and sadly, there weren't many. While most theologians were open to the possibility of God working in the wider world, even that there might be revelation in other religions, hardly any went the next step and attempted to explore that revelation. This pattern is, in part, the result of pressing cultural and historical issues (i.e. fashioning the church's dogma, reforming the Catholic church) that precluded theologians from pursuing their own conclusions about how insight from other religious faiths might be a part of God's general revelation to us. As we noted, our cultural and epistemological concerns are different, and so require a response from us in the form of pluralistic theology. However, when three contemporary examples of incorporation were examined to see what the process of incorporation might look like, a series of difficulties with such an approach became apparent. Hick's theology was appropriately self–titled as "christian" (with a small "c"), and as such did not bear much resemblance to the specificity of the

Christian tradition. Heim's work described how a pluralistic theology might look, but he never got around to examining the tenets of other religions. Here was the all–too–familiar pattern we observed in the history of the church. Finally, Smart's system fell prey to the same critique of Hick, namely, that it did not resemble Christianity anymore. Thus at the end of our investigation of Christian theology's used of "revealed" insight from other religious traditions as a resource for deepening self–consciously Christian thinking, we noted the lack of dialogue partners theologically. In short, there is a dearth of theologians to draw upon as we seek a pluralistic Christian theology of creation.

How then can we incorporate material from other religions into a Christian theology? Finding few theological companions from the Christian tradition, a model was found in the Old Testament in its wisdom literature. Here, Israel freely borrowed from her neighbors, even while crafting and shaping this material to suit her own needs. The method we observed was that Israel used material from other religions if it corresponded with (or did not overtly contradict) material present in the Israelite tradition, and if it provided a way to emphasize distinctive elements in the Israelite tradition. However, Israel went beyond looking for agreeable ideas – she learned from different ideas (i.e. Lady Wisdom). From the religious ideas embedded in other cultures, Israelite wisdom writers heard God speaking afresh – correcting, challenging, expanding. From Old Testament wisdom literature, we discovered a model for doing pluralistic Christian theology. This model includes discovery, locating similar and dissimilar themes with the Christian story, revisiting the theological guideposts for Christian theology in those areas, adapting the material to Christian doctrine, and finally presenting it back to the larger theological community for its edification.

With wisdom literature as our companion and guide methodologically, we then examined Buddhism, Hinduism, and Islam's cosmogonies and cosmologies. We found a number of areas of disagreement and a few areas of similarity in the areas of the nature of God, the nature of humanity, dualism, cosmology, cosmogony, eschatology, and immanence. However, it was on three areas of thought – preservation, transcendence, and stewardship of the cosmos that we chose to focus, for here we found both congruence and challenge to our own Christian thought.

We then explored these three areas as central mooring points in a pluralistic Christian doctrine of creation through a dialogue of Christian, Buddhist,

Hindu, and Islamic sources. In our reflection on preservation, we noted how the Buddhist ontology of co–dependence might challenge the individualism of much of Christian theology. There are resources within the Christian tradition here – notably in the notion of koinonia, and the role of the other – that we can draw upon, but these have not been emphasized nearly enough. Buddhist thought can provide an impetus, causing Christian theology to ask new questions and discover these more communal perspectives. The immanence of the Hindu deity Viṣṇu pushes us to re–emphasize the immanence present in the Christian Trinity, but even more centrally, the role of the Spirit. Pneumatology is not only a key element of Christian doctrine, but also the motivation, via the Spirit's role in immanence, in helping us to reach out, discern, and incorporate material from other religions. Islam's push for the "successive" new acts of God nudged us toward accepting the term "continuing creation," rather than "continued creation". In continuing creation we see the continuing work of God in fresh and new ways, which encourages us to seek the new, the developing, the evolving, the current, and the future of not only theology, but of God's working in the world.

In considering transcendence, we noted how the role of "emptiness" in Buddhism might stir us toward considering not only the "unknowability" of God, given the divine otherness, but also the lack of finality in any of our descriptions of God. A recovery of "kenosis" becomes key for the Christian doctrine of God at this point, a doctrine that will continue to be crucial if Christianity is to be a part of pluralistic theology in general. Hinduism's retention of transcendence as a key to its plurality of gods might similarly be instrumental in pushing Christian theologians to agree with Rahner and Barth in retaining transcendence in Trinitarian studies. It might also cause us to reject the lack of transcendence found in theology rooted in process thought. Interestingly, such study forced us to see the divine as both more deeply transcendent, as well as more deeply immanent. Islam, in this area, puts forth a comparatively less–developed role for the human in the face of divine transcendence than does Christian theology. Here we emphasized the work of Molina and Erasmus in articulating the freedom of the human alongside of divine transcendence. Retaining both of these emphases has been important in guarding against both a divine monism and a debilitating pantheism.

With regard to the stewardship of the cosmos, we invited a similar process through which we might learn from Hinduism and Buddhism in the area of environmental concerns. A focus on becoming stewards of creation in

Christian theology meant debating and correcting not only the misunderstandings of the past Christian tradition (which are, in some cases, some of the same issues facing the other traditions), but also being open to new areas of thought from other religious traditions. Buddhism's strong emphasis on the interconnectedness of all created beings and the ethical imperatives resulting from this forced us to consider "communal ethics" as a key to stewardship of creation. Hinduism's fluid epistemological categories for anthropology helped us to reconsider the importance of the Noahic covenant for all of creation, and not simply begin and end the hierarchy of creation with discussions of the *imago dei*. Humanity must also be seen as falling under a *theocentric* view of the universe, a role lost in perhaps a too modern conception of anthropology. Finally, Islam's strong role for humanity as "vice–regent" for God moves us to consider the kind of "dominion" humanity is meant to exercise over creation, one of service, love, and interplay. The Sabbath becomes key here as a way for the human and divine to enjoy creation together.

In all of these areas, it is the stimulus and insight from other religions that helps to set the agenda for Christian theology. Some of these emphases can be found present in the Christian tradition, but they might helpfully be brought to prominence given other theological misconceptions too often present. Others, however, are largely new. Our task in constructing a truly pluralistic Christian theology of creation has been to let these other religious traditions lead in crafting our theologies in the realm of general revelation. However, the theological process is not complete without a return to our tradition, to special revelation, and in concert with the findings from general revelation, to present our completed doctrine. This we have done in these three areas of preservation, transcendence, and stewardship of the cosmos. This last step both locates us squarely in our Christian tradition and allows us to be part of the wider interreligious dialogue.

But regarding our model, in referring to the theological guideposts found in the Christian tradition, it is clear we must consider both the "antecedal" and "non–antecedal" nature of pluralistic theology. Much of what the other traditions have to offer Christianity is already found in the Christian tradition in some form or fashion. These emphases are not wholly like their counterparts in other religions, nor do they possess the kind of embodied authenticity that would characterize them if seen in their own narratives and histories. But nevertheless many of them are present in the Christian tradition, both in the form of orthodoxy and heterodoxy. What that says to us is that the Christian tradition is wide and varied, and if we follow the methodology we have outlined above, we will usually find some kind of antecedent to any new incorporations we may wish to do. This "antecedent" step is important to maintain, I believe, if we are to remain as part of the Christian tradition in a "traditional" sense.

However, there is much more to this enterprise than maintaining the status quo, and here is where we must consider the "non–antecedal" nature of pluralistic theology. Christian theology must decide anew what is to be done with theology that pushes the boundaries of what is thought to be orthodoxy. If we are serious both about what our study has noted as the "continuing creation" of God, and of rediscovering the crucial pneumatological role in creation, we must remain open to the possibilities of new formulations of key doctrines. The "continuing creation" of God means new vistas in theology must remain open (in our case it is the arena of other religions), just as Israel was open to new ideas around her. But even more important is the role of the Spirit. The Spirit provides not only an immanent sense of God working in all of creation, but also the *theological grounding point* from which all pluralistic theologies can emerge. This emergence is not based, then, in an adherence to a worldview – i.e. history of religions, pluralism, etc. – but rather is grounded theologically in the third member of the Trinity. This theological starting point is the profound correction of Barth, and must guide us in this endeavor as well.

If other religions have potentially new understandings to impart to Christianity, we must decide what our hermeneutic is to be. Will it be one of incorporation or "learning from" the other? Or will it be one of rejection *in toto*? Learning from the other will require not simply revisiting old doctrinal battles for the present day, but keeping a conviction, a mindset, open toward the new. If the Spirit is the grounding point for all God's working in the world, we needn't be afraid of what God may be up to today. And here is where a sound natural theology and a generous doctrine of general revelation can pay wonderful dividends. If we have developed our postmodern theology of general revelation strongly enough, we can be open to the new leadings of God's Spirit.

The question may rightly be asked at this juncture, "Why?" Why not merely drink from the stream of revelatory water flowing through the Christian tradition, rather than concern ourselves with the "trickle" that may be flowing through other traditions? Calvin stated that not "one in a hundred" recognized God's working creation; if that is so, why bother? Again we must return to the Spirit. If the Spirit is "alive and well" around the world, working in creation and in the hearts of humanity, we do well to pay attention to it. The eschatological "new creation" is waiting for all of humankind, and none of us can be presumptuous as to its whole meaning.

This exploration of the "non–antecedal" in theology is merely pushing the envelope of the fourth corner of Wesley's quadrangle – experience, or what Robert Johnston has called "culture". Our culture, our experience today is one of plurality, diversity, and yes, difference. How can Christian theology make use of difference? Postmodernity has helped us epistemologically in this area, but has also pushed us too often to be content with diffusement and unconcerned singularity. We need to move toward incorporation under the aegis of the Spirit. Theologically, we have precious few resources of this type. It is my hope that this exploration, this constructive theology of creation, which makes use of the insights of the Spirit found in Buddhism, Hinduism, and Islam, might help in this new endeavor. This book is offered up to readers both as a methodological resource, and as a nascent study of a topic in systematic theology that may prove key for future pluralistic studies – creation.

Psalm 24:1 – "The earth is the Lord's and the fullness thereof, the world and those who dwell therein."

NOTES

Chapter One

[1] Dietrich Bonhoeffer, *Creation and Fall* (London: SCM Press, 1959), pp. 9–10.

[2] The literature on "postmodernity" is voluminous, and space does not permit us to delineate the multi–faceted aspects to this movement. For the sake of this book, we will assume that the transition from modernity to postmodernity has occurred, though this assumption is not necessary for us to proceed to examine the doctrine of creation and pluralism. Hal Knight has suggested five modifications for postmodernity *vis a vis* modernity that are helpful for us to consider as we continue: 1) From Individualism to Community; 2) From Rational Foundationalism to Nonfoundationalism; 3) From Methodological Doubt to Traditioned Belief; 4) From Dualism to Holism; 5) From Optimism to Pessimism. Cf. *A Future for Truth* (Nashville: Abingdon, 1997), pp. 53–56. Many of these transitions hold much promise for contemporary theologians, in my judgment, though perhaps a better rendering of point #5 would be from Optimism to Paradox.

[3] Ludwig Feuerbach, *The Essence of Christianity*, trans. George Eliot (New York: Harper and Brothers, 1957), p. 287.

[4] Emil Brunner may believe that this is possible, even though he himself does not attempt it. He writes that the creation accounts of Genesis don't offer a specifically Christian account of creation – one needs to proceed to the Fourth Gospel and the Epistles to do that; and he criticizes Barth for remaining in the Old Testament in his exposition. Emil Brunner, *The Christian Doctrine of Creation and Redemption – Dogmatics, vol. II*, trans. Olive Wyon (London: Lutterworth Press, 1952), p. 7.

[5] Paul Knitter, "Introduction", *The Myth of Christian Uniqueness – Toward a Pluralistic Theology of Religions*, ed. John Hick & Paul F. Knitter (Maryknoll, NY: Orbis, 1987), p. vii.

[6] Paul Knitter, "Introduction", *The Myth of Christian Uniqueness – Toward a Pluralistic Theology of Religions*, p. viii. . Harold Netland feels that the term *exclusivism* "was introduced into the discussion not by adherents of the traditional perspective but rather by those who rejected this view and wished to cast it in a negative light. It is a pejorative term with unflattering connotations ..." (*Encountering Religious Pluralism* (Downers Grove, Il: Intervarsity Press, 2001), p. 46). Netland prefers the term *particularism* as a less–offensive label: "Contrary to popular assumptions, there is no reason to suppose that particularists cannot be culturally sensitive or appropriately tolerant and accepting of followers of other religious traditions... Particularism does not entail that all the beliefs of non–Christian religions are false. It does not imply that there is nothing of value in

other religions, nor that Christians cannot learn anything from other religions" (pp. 48–49).

[7] See for example Kärkkäinen's four categories: 1) Ecclesiocentrism – "the exclusive attitude, according to which religions are not salvific or even necessarily condusive to the search for God, and salvation can be found only in the Christian church, the locus of faith in Christ;" 2) Christocentrism – "the inclusive approach, according to which Christ is the Savior but the benefits of his saving work may be found outside the Christian church and Christian religion;" 3) Theocentrism – "the pluralistic paradigm, according to which Christ is one savior among other savior figures and not an exclusive one. In this view, God alone stands at the center. The various religions, Christianity included, represent many ways leading to God;" 4) Realitycentrism – "the center of religions is not a God or gods but an ultimate reality." Cf. Veli–Matti Kärkkäinen, *An Introduction to the Theology of Religions* (Downers Grove: InterVarsity Press, 2003), p. 25. Cf. also J. Dupuis, *Toward a Christian Theology of Religious Pluralism* (Maryknoll, NY: Orbis, 1997), pp. 180ff. In Knitter's latest work, *Introducing Theologies of Religions* (2003), he suggests four ways of conceiving of pluralism: The Replacement Model ("only one true religion"); The Fullfillment Model ("the one fulfills the many"); The Mutuality Model ("many true religions called to dialogue"); The Acceptance Model ("many true religions: so be it"). Cf. Paul Knitter, *Introducing Theologies of Religions* (Maryknoll, NY: Orbis, 2003).

[8] Knitter, *The Myth of Christian Uniqueness*, ed. Hick & Knitter, p. ix.

[9] John Hick, "The Non–Absoluteness of Christianity", *The Myth of Christian Uniqueness – Toward a Pluralistic Theology of Religions*, ed. John Hick & Paul F. Knitter (Maryknoll, NY: Orbis, 1987), p. 30.

[10] Knitter, p. x.

[11] Hans Kung, "Christianity and World Religions: Dialogue with Islam", *Toward a Universal Theology of Religion,* ed. Leonard Swidler (Maryknoll, NY: Orbis, 1987), p. 194.

[12] Raimundo Panikkar, "The Jordan, the Tiber, and the Ganges", *The Myth of Christian Uniqueness – Toward a Pluralistic Theology of Religions*, ed. John Hick & Paul F. Knitter (Maryknoll, NY: Orbis, 1987), pp. 89–90.

[13] Panikkar, "The Jordan, the Tiber, and the Ganges", ed. Hick & Knitter, p. 90.

[14] Christoph Schwöbel, "Particularity, Universality, and the Religions", *Christian Uniqueness Reconsidered*, ed. Gavin d'Costa (Maryknoll, NY: Orbis, 1990), p. 33.

[15] John B. Cobb, Jr., "Beyond 'Pluralism'", *Christian Uniqueness Reconsidered*, ed. d'Costa, p. 81.

[16] Cobb, p. 82.

[17] See also Gordon D. Kaufman, "Religious Diversity and Religious Truth" in *God, Truth and Reality: Essays in Honor of John Hick*, ed. Arvind Sharma (New York: St. Martin's Press, 1993); and John Milbank, "The End of Dialogue," in *Christian Uniqueness Reconsidered*, ed. d'Costa.

[18] John Hick, *An Interpretation of Religion* (New York: Macmillan, 1989), pp. 30ff.

[19] Hick, *An Interpretation of Religion*, pp. 325ff.

[20] Ninian Smart & Steven Konstantine, *Christian Systematic Theology in a World Context* (Minneapolis: Fortress, 1991), p. 91.

[21] Gavin d'Costa, "Preface" in *Christian Uniqueness Reconsidered*, ed. d'Costa, p. ix.

[22] Kathryn Tanner, "Respect for Other Religions: A Christian Antidote to Colonialist Discourse," *Modern Theology* 9 (1993): 1.

[23] Kevin J. Vanhoozer, "Does the Trinity Belong in a Theology of Religions?", in *The Trinity in a Pluralistic Age*, ed. Vanhoozer (Grand Rapids: Eerdmans, 1997), p. 68.

[24] John Hick, *A Christian Theology of Religions* (Louisville: Westminster John Knox Press, 1995), pp. 40–41. Quoting Kenneth Surin, "Towards a 'Materialist' Critique of 'Religious Pluralism', *The Thomist*, Vol. 53, no. 4, October 1989, p. 666.

[25] Hick, *A Christian Theology of Religions*, p. 41.

[26] Ibid., p. 41.

[27] Ibid., p. 43.

[28] Ibid., p. 42.

[29] See below, pp. 97ff.

[30] Dupuis, *Toward a Christian Theology of Religious Pluralism*, p. 6.

[31] Dupuis, p. 7.

[32] Ibid., p. 7.

[33] Kärkkäinen, *An Introduction to the Theology of Religion*, p. 21.

[34] While we do not have the space to devote to this area, a few points are worth mentioning: 1) Biblical creation stories, in contrast to Babylonian and Egyptian cosmogonies, bring together both universal creation, and the creation of the human; 2) Genesis lays unusual stress on the role of the human; 3) There is a fundamental demarcation between the divine and human spheres in the Biblical story, as opposed to the variety of options in Egyptian and Mesopotamian cosmogonies; 4) Though creation by the word was not unusual in the Ancient Near East, Israel's conception was informed by her prophetic tradition rather than her surrounding cosmogonies. See Claus Westermann, *Genesis* (Minneapolis: Augsburg Fortress, 1992); Gerhard Von Rad, *Old Testament Theology*, vol. 1 (New York: Harper & Row, 1962); *Creation in the Biblical Traditions*, ed. Richard J. Clifford & John J. Collins (The Catholic Biblical Quarterly Monograph Series, 24, 1992).

[35] In a similar methodological approach, Niebuhr rejects both the "Christ or Culture" position and the "Christ against Culture" perspective, believing that an adequate theology must bring both "Christ and Culture" into an ongoing relationship. Cf. H. Richard Niebuhr, *Christ and Culture* (New York: Harper & Row, 1951).

Chapter Two

[1] Lyotard writes, "postmodernism is not modernism at its end, but in a nascent state, and this state is recurrent." Cf. *The Postmodern Explained*, trans. Don Barry (Minneapolis, 1993), p. 13.

[2] See Gregory Bruce Smith's *Nietzsche, Heidegger and the Transition to Postmodernity* (Chicago: University of Chicago Press, 1996), pp. 1–9.

[3] R. Pippen, *Modernism as Philosophical Problem* (Cambridge Mass.: Basil Blackwell, 1991), p. 156.

[4] Smith, p. 8. Smith also writes, "Throughout postmodernism there is a fascination with the surface, conjoined with a perception that all of existence is a humanly constructed

surface that has about it the aura of 'contingency'. . . This understanding leads to an ironic attitude toward life that inevitably transforms itself into a form of cynicism . . . ." (pp. 9–10). This attitude has of course had philosophical, literary, and architectural manifestations in our time. A more positive viewpoint of postmodernity would be to see it not promoting "no hope", but instead offering a critique of false hope.

5 Os Guiness, *Fit Bodies, Fat Minds* (Grand Rapids: Baker, 1994), p. 106.

6 See J. Milbank's critiques in *Theology and Social Theory* (Oxford, 1990), pp. 12–17; 27–45.

7 Nancey Murphy, *Beyond Liberalism and Fundamentalism: How Modern and Postmodern Philosophy Set the Theological Agenda* (Valley Forge, PA: Trinity Press International, 1996), p. 4.

8 Renè Descartes, *Discourse on Method* (1637), part II.

9 Murphy, *Beyond Liberalism and Fundamentalism*, pp. 12–13. See also Nicholas Wolterstorff, *Reason Within the Bounds of Religion* (Grand Rapids: Eerdmans, 1976).

10 Murphy, *Beyond Liberalism and Fundamentalism*, p. 12

11 Carl F.H. Henry, *God, Revelation and Authority* (Waco, TX: Word, 1976–1983), vol. 3, p. 457.

12 Donald G. Bloesch, *Holy Scripture: Revelation, Inspiration and Interpretation* (Downers Grove, IL: Intervarsity Press, 1994), p. 20.

13 cf. "We have no formula for the being of God in itself as distinct from the being of God in the world" in *The Christian Faith* (Edinburgh, 1928), p. 172. This sense of immanence will become an important topic for us later (cf. pp. 217 ff.)

14 David Tracy, *The Analogical Imagination* (New York: Crossroad, 1981), p. 113.

15 Stephen Toulmin, *Cosmopolis: The Hidden Agenda of Modernity* (New York: Macmillan, 1990), ch. 1.

16 cf. David Tracy's *Blessed Rage for Order: The New Pluralism in Theology* (New York: Seabury, 1978). Tracy writes in another context that in doing theology, one must recognize "a radical *pluralism*, {italics mine} indeed, even an intense conflict of interpretation, from which there can often seem no honorable exit". David Tracy, "Theological Method," in *Christian Theology: An Introduction to Its Traditions and Tasks*, 2nd ed., ed. Peter C. Hodgson & Robert H. King (Philadelphia: Fortress, 1985), p. 37.

17 Graham Ward, "Postmodern Theology" in *The Modern Theologians*, 2nd ed., ed. David F. Ford (Cambridge, Mass: Blackwell, 1997), p. 585.

18 Hans Georg Gadamer, *Truth and Method* (New York: Crossroad, 1982), p. 235.

19 Gadamer, *Truth and Method*, p. 173–74.

20 see Murphy, pp. 91ff.

21 Jeffrey Stout, *The Flight from Authority* (Notre Dame: 1981), p. 6.

22 An interesting point to discuss here is whether a postmodern is always a post– or nonfoundationalist. J. Wentzel van Huyssteen argues no. In responding to Nancy Murphy's adoption of Imre Lakatos' scientific work, he writes, "Is the postmodernist, then always a postfoundationalist? In theology, at least, this often hardly seems to be the case. In fact, the postmodernist might even turn out to be a crypto–fideist", in *Essays in Postfoundationalist Theology* (Grand Rapids: Eerdmans, 1997), p. 90. For instance, van Huyssteen argues that Murphy's incorporation of Lakatos' "hard core beliefs" into her

system – i.e. the existence of God is assumed – is possibly a "surreptitious form of foundationalism" (pp. 87–90). Compare this with John E. Thiel's statement: "I would argue, though, that nonfoundational theology affirms the most basic traits that modern theologies share and, for that reason, would understand the prefixes in "postliberal" and "nonfoundational" approaches to theology to be expressions of dissatisfaction with, but not rejections of, modern assumptions" in *Nonfoundationalism* (Minneapolis: Fortress, 1994), p. 115. My own answer is that I believe non– or post–foundationalism to be at the very center of what it means to be postmodern. In fact, to be a foundationalist is to be modern.

23 In *From a Logical Point of View* (Cambridge: Harvard University Press, 1953).

24 Quine, *From a Logical Point of View*, pp. 42–43.

25 Murphy, p. 94.

26 cf. see Carl F.H. Henry's article, "Postmodernism: The New Spectre", in *The Challenge of Postmodernism: An Evangelical Engagement* (Wheaton: Victor Books, 1995), ed. David S. Dockery, pp. 34–52.

27 Alasdair MacIntyre, *Three Moral Versions of Enquiry* (Notre Dame: University of Notre Dame Press, 1989), p. 5.

28 Cf. George Lindbeck's doctrine as grammar in *The Nature of Doctrine* (Philadelphia: Westminster Press, 1984).

29 Elizabeth Barnes, "Community, Narrative, and an Ecological Doctrine of Creation" in *Theology without Foundations: Religious Practice and the Future of Theological Truth*, ed. Stanley Hauerwas, Nancey Murphy, Mark Nation (Nashville: Abingdon, 1994), pp. 181ff. Cf. Murphy, p. 27.

30 Terrance Tilley, "In Favor of a 'Practical Theory of Religion'", in *Theology without Foundations: Religious Practice and the Future of Theological Truth*, ed. Stanley Hauerwas, Nancey Murphy, Mark Nation (Nashville: Abingdon, 1994), pp. 49ff.

31 Trevor Hart, *Faith Thinking* (Downers Grove, Ill.: Intervarsity Press, 1995), pp. 101–102.

32 Colin Brown, "Natural Theology", in *New Dictionary of Theology*, ed. Sinclair B. Ferguson & David F. Wright (Downers Grove, Ill: InterVarsity Press, 1988), p. 452.

33 Donald McKim, *Westminster Dictionary of Theological Terms* (Louisville: Westminster John Knox Press, 1996), p. 240.

34 Aquinas, *Summa Theologica*, 1a 2,2. Cf. *The Summa Theologica of St. Thomas Aquinas* (New York: Benziger Bros., 1947). Aquinas explicates this further in elucidating the uses of philosophy: First, "we can use it to demonstrate the preambles of faith … for example, that God exists, that God is one, or similar propositions concerning God or creatures that faith proposes as having been proved in philosophy." Second, "we can use philosophy to make known through certain likenesses what belongs to faith, as Augustine in his book *On the Trinity* uses many likenesses drawn from the teachings of the philosophers to explain the Trinity. Third, we can use philosophy to oppose what is said against faith, either by showing that these things are false or by showing that they are not necessary" (*Summa contra Gentiles,* I, 9). Cf. *Summa Contra Gentiles* (Notre Dame: University of Notre Dame Press, 1975).

35 Aquinas, *De Veritate*, XIV, 10, reply. Cf. *De Veritate* (Chicago : H. Regnery Co., 1952–54).

36 *Summa Theologica*, 2a2ae. 2, 4.

[37] *Summa contra Gentiles*, I, 7, [1].

[38] Arvin Vos, *Aquinas, Calvin & Contemporary Protestant Thought* (Washington, D.C.: Christian University Press, 1985), pp. 111–112.

[39] See Vos, pp. 111–112.

[40] Building on the work of G. de Broglie, Vos argues that "the Thomists were faithful to Aquinas's conception of the preambles of faith through the sixteenth century, but that in the middle of the seventeenth century, especially under the influence of Cartesianism, they began to be concerned with presenting the act of believing as a rationally viable act. In this context natural theology took on a whole new role..." (Vos, p. 115). See also G. de Broglie, "La vraie notion thomiste des 'praeambula fidei,' " *Gregorianum* 34 (1953) : 341–389.

[41] In fact, in a fascinating passage, Aquinas notes, "Thus, against the Jews we are able to argue by means of the Old Testament, while against heretics we are able to argue by means of the New Testament. But Mohammedans and the pagans accept neither the one nor the other... we must, therefore, have recourse to the natural reason, to which all men are forced to give their assent" (*Summa theologiae*, 1a. 2, 2.).

[42] Commentary on Psalm 104:1, *Corpus Reformatorum*, ed. Baum, Cunitz, & Reuss (Brunswick, 1863–1900); vol. 32:85. Quoted in Ronald Wallace, *Calvin's Doctrine of the Word and Sacrament* (Grand Rapids: Eerdmans, 1957), p. 66.

[43] John Calvin, *Institutes of the Christian Religion.* Tr. By Ford Lewis Battles. Ed. by John T. McNeill. Library of Christian Classics. Vol. 20 (Philadelphia: Westminster Press, 1960), I.14.21

[44] Calvin, *Institutes of the Christian Religion*, I.5.1.

[45] T.H.L. Parker, *Calvin's Doctrine of the Knowledge of God* (Grand Rapids: Eerdmans, 1959), p. 18.

[46] Wallace, p. 66. For Calvin, "in the time of Cain, Abel and Noah there was a universal religious tradition (cf. comm.. on Lev. 11:2, *Corpus Reformatorum* 24:345f.: on Gen. 4:2f, *C.R.* 23:83f; on Gen. 12:7, *C.R.* 23:180f.). This was transmitted mainly through Shem (comm.. on Gen. 33:20, *C.R.* 23:453), but adulterated forms of true ritual descended to the Gentiles and affected their rites (cf. on Exod. 29:38, *C.R.* 24:491; on Gen. 35:17 fl, *C.R.* 23:473)" (Wallace, p. 66).

[47] Calvin, *Institutes*, I.3.1.

[48] Ibid., I.5.14.

[49] Wallace, p. 67. Wallace continues, "'We know the profane to be endowed with singular talents, and other eminent faculties; and these are called natural, since God desires by His gracious gifts to shine forth in the human race by such examples as these.' But when it comes to the knowledge of God such gifts are excluded from consideration as a means of achievement. 'Daniel places revelation on a higher footing.' 'Art and science cannot obtain the power of understanding the revelation of God, but a revelation from the Spirit must be waited for'" (Wallace, pp. 67–68).

[50] Calvin, *Institutes*, I.4.1.

[51] Ibid., I.4.4. See further Calvin's comments on how the imago Dei has been destroyed or corrupted by sin. *Institutes*, II.2.12.

[52] Susan Schreiner argues that though Calvin opposed natural theology, he favored a theology of nature. Cf. *The Theatre of His Glory* (Grand Rapids: Eerdmans, 1995).

[53] William J. Bouwsma, *John Calvin: A Sixteenth Century Portrait* (New York: Oxford University Press, 1988), pp. 3, 13.

[54] Bouwsma, pp. 14, 17.

[55] McDermott perhaps overstates his case when he remarks, "Calvin is yet another major Christian theologian who was helped by a pagan thinker to understand better the nature or meaning of Christian revelation." Gerald R. McDermott, *Can Evangelicals Learn from World Religions?* (Downers Grove, Il: Intervarsity Press, 2000), p. 132. Calvin certainly would not have characterized his thoughts on Christian revelation as being influenced by Cicero or Quintilian – Christian revelation can only come through Christ. It is perhaps in the organization of Calvin's work that his indebtedness to the humanists is most present.

[56] Emil Brunner, in *Natural Theology* (London: The Centenary Press, 1945), pp. 25–26.

[57] Brunner, *Natural Theology*, p. 31.

[58] Brunner even cites a passage where "Calvin leaves room for the thought that some few pious men among the heathen might have known the true God and might thus have attained to that righteousness which counts before God" (p. 44). Cf. *C.R.*, vol. 33, p. 27; Vol. 25, p. 266. Brunner says the fault of Barth's appropriation of the Reformers for his anti–natural theology viewpoint is in a misunderstanding of the concept of nature: "But the fault of this lies above all in the fact that the concept of nature, which was common to all theology until the time of the Enlightenment, had been lost and was therefore not understood by the theologians of the nineteenth century who went back to the Reformers. It need hardly be said that the concept of nature which gained currency at the time of the Enlightenment was completely different both from that of Catholicism and from that of the Reformers. If one wanted to put it metaphorically, one would have to say that for the Reformers the light of the revelation in Christ must shine into nature in order to light up this foundation. The Roman Catholics separate them by a neat horizontal line. For the Enlightenment the light of reason reaches upwards into the sphere of redemption to the extent of doing away completely with the distinction between the *lumen naturale* and the revelation in Christ. The *lumen naturale* now becomes itself a revelation, indeed the only one that there really is. Thus the whole of theology becomes a *theologia naturalis*" (p. 47).

[59] Karl Barth, in *Natural Theology*, pp. 74–75.

[60] Barth, in *Natural Theology*, pp. 91–92.

[61] Ibid., pp. 105–107.

[62] Ibid., p. 116, 120, & 121.

[63] Ray Anderson points out some wrinkles in Barth's thinking in this area, particularly in Barth's desire to preserve a correspondence between knowledge of God in creation and knowledge of God through Christ. See "Barth and a New Direction for Natural Theology," *Theology Beyond Christendom*, Ed. John Thompson (Allison Park, PA: Pickwick, 1986), pp. 241–266. While Anderson sees Barth as continuing his denial of an "independent" natural theology, and that Barth's Trinitarian thrust enabled him to sever the link between humanity and deity, Anderson nevertheless suggests that Barth was open to a contingent relation between God and the created world through a unitary view of creation and incarnation.

[64] Wolfhart Pannenberg, *Systematic Theology*, vol. 1, trans. G. Bromiley (Grand Rapids: Eerdmans, 1991), pp. 73–74.

[65] H. Denzinger, *The Sources of Catholic Dogma*, tr. R.J. Defarrari (from 30th ed.; St. Louis: Herder, 1957), p. 443.

[66] The result being, according to O. Weber: "To leave 'the natural' without reservations and without theological interest to the natural sciences and instead of it to discuss God solely in the context of the moral life." Weber then outlines a fascinating development of the doctrine: "We see in both Ritschl and Herrmann that the exclusion of nature from the realm of theological interest had profound results within the area of what once was called special revelation, especially in Christology. For Ritschl, Christological statements were reduced to an ethically determined 'value judgment,' and for Herrmann they were only related to the 'inner life' of Jesus as the locus of revelation. Therefore it ought not to be surprising that the dogmaticians of that period who came out of Biblicism, especially Martin Kähler, Adolf Schlatter, and more recently, Paul Althaus, show a tendency, sometimes cautiously (Kähler), sometimes radically (Schlatter), sometimes critically, not to restrict the knowledge of God exclusively to the realm of inwardness and the ethical. As a result, they end up speaking of a preparatory stage to the knowledge of God based upon the Bible, a stage established upon the perception of self and the world. The very emphatic stress placed upon general revelation and common grace in Kuyper and his associates is also a result of their opposition to ethicism." O. Weber, *Foundations of Dogmatics*, vol. 1, trans. D. Guder (Grand Rapids: Eerdmans, 1981), pp. 203–204.

[67] Pannenberg, p. 102. Pannenberg sees the roots of this in a faulty understanding of Feuerbach (pp. 104ff.).

[68] Pannenberg notes, "For all our criticism of the natural theology of the Baroque age and the Enlightenment, we must acknowledge that its argument in fact was the very opposite, seeking to show that the existence of humanity and its world would not be possible without the existence of God" (p. 106).

[69] Ibid., pp. 107–108.

[70] Both Barth and Brunner went on to modify their positions somewhat, according to Ray Anderson. In *Man in Revolt*, Brunner suggests that the notion of the *imago* is to be understood as a dialectical relation between the *imago*–origin and the humanity–of–the–sinner (*Man in Revolt* {Philadelphia: Westminster 1962}, pp. 513–514). Barth, while following Luther in suggesting a total loss of the *imago* through the Fall, went on to posit that humanity could not lose the *imago* because it never possessed it in the first place {*Church Dogmatics*, III/1, (Edinburgh, T. & T. Clark, 1975), p. 200}. Cf. Ray Anderson, *On Being Human* (Grand Rapids: Eerdmans, 1982), pp. 221ff..

[71] Calvin, *Institutes of the Christian Religion*, 2.2.15.

[72] Amos Yong, *Beyond the Impasse* (Grand Rapids: Baker, 2003), p. 21.

[73] Jürgen Moltmann, *God in Creation* (Minneapolis: Fortress, 1993), p. 99.

[74] Moltmann, *God in Creation*, p. 100.

[75] Colin Gunton, "The Trinity, Natural Theology, and a Theology of Nature," in *The Trinity in a Pluralistic Age*, ed. Kevin J. Vanhoozer (Grand Rapids: Eerdmans, 1997), pp. 102–103. Here I would add, in light of our subject, God's beautiful creativity found in other religions can be as much of a focal point of religious study as good art, psychology, architecture, etc.

[76] Pannenberg, pp. 117–118.

[77] Moltmann, *God in Creation*, p. 59.

[78] Ibid., p. 59.

[79] Ibid., p. 60.

[80] Ibid., p. 60.

[81] Another possible theological starting point for integrating natural and revealed theology may be "common grace". Cf. Robert Johnston, "Rethinking Common Grace", in *Grace Upon Grace*, ed. Robert Johnston, L. Gregory Jones, & Jonathon R. Wilson (Nashville: Abingdon, 1999).

[82] Colin Gunton, "The Trinity, Natural Theology, and a Theology of Nature," p. 102.

Chapter Three

[1] Colin Gunton, *The Triune Creator* (Grand Rapids: Eerdmans, 1998), p. 10.

[2] Colin Gunton, *The One, the Three, and the Many: God, Creation and the Culture of Modernity* (Cambridge: Cambridge University Press, 1993).

[3] Trinitarian epistemology, particularly relating to pluralism, has been an oft–explored theme. See section on Hinduism below, pp. 180ff.

[4] Colin Gunton, "The Trinity, Natural Theology, and a Theology of Nature," p. 103.

[5] Gunton, "The Trinity, Natural Theology, and a Theology of Nature", p. 98.

[6] Ibid., p. 101.

[7] Gunton, *The Triune Creator*, p. 103.

[8] Ibid., p. 192.

[9] Gunton, *The One, The Three, and the Many*, pp. 180.

[10] Gunton, *The Triune Creator*, p. 37.

[11] Ibid., p. 39. Gunton asserts that in fact the doctrine of creation, in its implied critique of Aristotle, allowed for science and theology to coexist together: "It was, especially in making the idea of the rational contingency of a world distinct from God, undoubtedly a necessary although not sufficient condition for the rise of science" (p. 116).

[12] Cf. Ibid., p. 8.

[13] Ibid., p. 83.

[14] Ibid., p. 54.

[15] Ibid., p. 154.

[16] Colin Gunton, "The Spirit Moved Over the Face of the Waters: The Holy Spirit and the Created Order." *International Journal of Systematic Theology*, vol. 4, 2, July 2002, p. 195.

[17] Gunton, "The Spirit Moved Over the Face of the Waters," p. 195.

[18] Smart & Constantine, *Christian Systematic Theology in a World Context,* p. 22.

[19] Smart & Constantine, p. 441.

[20] Ibid., p. 21.

[21] Ibid., p. 201.

[22] Ibid., p. 209.

[23] Moltmann, *God in Creation*, p. 244.

[24] Smart & Constantine, p. 237.

[25] John Hick, *God Has Many Names* (Philadelphia: Westminster Press, 1982).

[26] Smart & Constantine, p. 217.

[27] Ibid., p. 235. This, we will see, is a correct implication of a rightly ordered stewardship of creation, pp. 247ff.

[28] Ibid., p. 244.

[29] We shall also note later *how* Smart makes use of his different sources. See pp. 105ff.

[30] Some may argue that Gunton's particularist approach makes it *more* accessible to a wider audience precisely because it is consciously Christian. It is, in effect, a true representation of Christianity to be presented for dialogue. However, Gunton's approach does not feed from as many worldwide religious streams as Smart's does, and, as such, is not positioned as a dialogue partner with other religions. Dialogue, at its core, involves an openness to the faith of the other in listening and learning. This attitude is not seen in theologies such as Gunton's. This key posture must be adopted, I believe, in this global, postmodern context for Christian theology to be relevant in the twenty–first century.

Chapter Four

[1] Justin, *Apology*, II, 5. In *The Ante–Nicene Fathers*, Vol. I, Ed. A. Roberts & J. Donaldson (Buffalo: The Christian Literature Pub. Co., 1885).

[2] Clement of Alexandria, *Stromateis*, I, 5. Quoted in *Attitudes Toward Other Religions*, ed. Owen C. Thomas (New York: Harper & Row, 1969), p. 16.

[3] Gerald Bray, "Explaining Christianity to Pagans", in *The Trinity in a Pluralistic Age*, ed. Kevin J Vanhoozer, (Eerdmans, 1997), pp. 24–25.

[4] Kenneth Scott Latourette, *A History of Christianity*, vol. 1, (New York: Harper & Row, 1975), p. 97.

[5] Eugene Portalie, *A Guide to the Thought of Saint Augustine* (London: Burns & Oates, 1960), p. 87.

[6] Portalie includes an intriguing section regarding Augustine's use of Neoplatonism delineating: 1) Christian theories wrongly thought to be platonic; 2) Platonic theories always approved for use in dogma; 3) Neoplatonic theories always rejected; and 4) Neoplatonic theories first adopted, then rejected (pp. 98–104).

[7] Augustine, *Against the Academics*, 20,43. Cf. *Ancient Christian Writers*, ed. J. Quasten & J. Plumpe (Westminster, MD: The Newman Press, 1950), vol. 12.

[8] Augustine, *On Christian Doctrine*, II, 40. Cf. *A Select Library of the Nicene and Post–Nicene Fathers of the Christian Church*, ed. P. Schaff, vol. II (Grand Rapids, Eerdmans, 1956).

[9] Augustine, *The Correction of the Donatists*, I, 2. Cf. *A Select Library of the Nicene and Post–Nicene Fathers of the Christian Church*, ed. P. Schaff, vol. IV.

[10] Augustine, *The Correction of the Donatists*, X, 46.

[11] Aquinas, *Summa Theologica*, IIa. 11.

[12] Aquinas, *Summa Theologica*, IIa.10. 8.

[13] Ibid., IIa. 10. 9.

[14] Ibid., IIa. 10. 11.

[15] Aquinas, *Commentary on the De Trinitate of Boethius*, II, a. 3. Cf. *Faith, Reason and Theology: questions I–IV of his Commentary on the De Trinitate of Boethius/ St. Thomas Aquinas*, trans. A. Maurer (Toronto: Pontifical Institute of Medieval Studies, 1987).

[16] Latourette writes, "the task to which Aquinas set himself was the reconciliation of Aristotle with the Christian faith" (vol. 1, p. 511).

[17] Gerald R. McDermott, *Can Evangelicals Learn from World Religions?*, p. 125. Cf. Timothy McDermott, *Thomas Aquinas: Selected Philosophical Writings* (New York: Oxford University Press, 1993), p. xxii.

[18] Brian Davies, *The Thought of Thomas Aquinas* (Oxford: Clarendon, 1992), p. 70, n. 38.

[19] Vos, p. 166.

[20] Less debated is the reason *why* Aquinas used so much Aristotelianism. A.C. Pegis offers this apologetical explanation: "St. Thomas himself may very well have thought that the *Summa Contra Gentiles* was precisely the sort of work needed by Christian missionaries in Spain face to face with the high intellectual culture of the Moslem world. Seen from this point of view, the *Summa Contra Gentiles* is a manual of apologetics against the intellectual picture of the universe created for the Western world by the translation of the writings of Aristotle and his followers into Latin in the course of the 12th and 13th centuries. In a large sense, therefore, the *Summa Contra Gentiles* is part of the Christian intellectual reaction against Arabian intellectual culture, and especially against Arabian Aristotelianism." Cf. Anton C. Pegis, "Introduction", in *Summa Contra Gentiles*, I (Notre Dame: University of Notre Dame Press, 1975), p. 21.

[21] *Luther's Works*, ed. J. Pelikan, vol. 23 (Saint Louis: Concordia, 1959), p. 82.

[22] *Luther's Works*, vol. 22, p. 60.

[23] This is not to say that Luther was completely unaware of the wider world. In fact, Pelikan believes his inclusion of Turks in his admonitions reflects his awareness of the interest that had developed among Christians because of renewed contact with Arabs and Turks (*Luther's Works*, vol. 6, p. 105). We must also mention Luther's usage of Occam (c. 1285 – 1349) and his nominalism. Nominalism taught that reason is to be subordinated to experienced reality; in revelation, scripture is the sole foundation – reason and experience do not precede but follow scripture. Luther's usage of Occam is an example of his usage of a source outside of specific revelation, but this source must be contextualized as a Christian theological/philosophical source as opposed to an Aristotle, whose secular philosophy captivated Aquinas. See Heiko A. Oberman, *Luther: Man between God and the Devil* (New Haven: Yale University Press, 1989), pp. 117ff.

[24] John Calvin, *Institutes of the Christian Religion*, 6:4.

[25] *Institutes*, 6 :1.

[26] Though we must note Calvin's usage of Cicero, which we have commented on earlier (cf. pp. 37ff.).

[27] F. Schleiermacher, *On Religion: Speeches to Its Cultured Despisers* (3rd ed.), trans. J. Oman (New York: Harper, 1958), p. 217.

[28] Schleiermacher, *On Religion*, p. 241.

[29] Owen Thomas, *Attitudes Toward Other Religions*, pp. 19–28.

[30] Here we would note Karl Barth's radical objection to this type of approach. Barth famously contradicts Schleiermacher's approach to religion particularly in his development of trinitarian doctrine. Barth argued that the "doctrine of the Trinity must be "decisive and

controlling for the whole of dogmatics", (*C/D*, I/1, p. 303) otherwise we run the risk of having "God as a general concept within which the Christian God as he is basically known in the doctrine of the Trinity is only a special instance" (Barth, *The Gottingen Dogmatics*, ed. Hannelotte Reiffen, trans. G. Bromiley (Grand Rapids: Eerdmans, 1991), pp. 97–98). Schleiermacher of course places the doctrine of the trinity at the end of his dogmatics, after he has defined the essence of religion, and as such it is a kind of summary statement, "not an immediate utterance concerning the Christian self–consciousness" (F. Schleiermacher, *The Christian Faith*, p. 738). This placement is misguided, according to Barth, for the doctrine of the trinity is not the conclusion to an examination of the attributes of God, but rather the presupposition to an explanation of God.

[31] Thomas himself argues differently, that in fact, religion is the foundation of culture. Cf. *Attitudes Toward Other Religions*, pp. 20ff.

[32] Ernst Troeltsch, *Christian Thought: Its History and Application* (London: University of London Press, 1923), p. 14.

[33] Troeltsch, p. 14.

[34] Ibid., p. 22.

[35] Ibid., p. 27.

[36] James Luther Adams, "Introduction", in Ernst Troeltsch's *The Absoluteness of Christianity and the History of Religions* (Richmond: John Knox Press, 1971), p. 17.

[37] Troeltsch, p. 34.

[38] Karl Barth, *Church Dogmatics*, 1/II, pp. 297–8, 199, 301, 302, 308, 326.

[39] We should note some further re–examination of Barth. Paul S. Chung looks at some of the expansive possibilities of Barth's Christology: "Barth's concern is to dialectically combine the word of Jesus Christ with various claims to truth in a pluralistic society. According to him, in the world reconciled with God in Jesus Christ, God cannot abandon any secular sphere. Even from the mouth of Balaam we recognize the well–known voice of the Good Shepherd, which should not be ignored despite its sinister origin. Forms of secularism or pluralistic claims to truth ought to be regarded as signs and guideposts pointing to the Word of Christ to whom other cultures and religions are enhypostatically related." Cf. Paul S. Chung, *Martin Luther and Buddhism (*Eugene, OR: Wipf & Stock, 2002). Cf. also Kärkkäinen, *Trinity and Religious Pluralism* (Ashgate, 2004), pp. 19ff. Barth's discussion of Mozart and his music applies here as well, Mozart having a great effect on the true Christian and on his knowing something about creation that the church fathers didn't, even though Mozart was not a "particularly active Christian" (*CD*, III/3).

[40] Barth, CD, III/1, pp. 80–82. "History" for Barth "is the history which is accessible to man because it is visible and perceptible to him and can be comprehended as history, is from the objective standpoint creaturely history in the content of other creaturely history, as an event prior to which and side by side with which there are other events of the same basic type with which it can be compared and integrated" (p. 78). Saga is a form of history – "history … recognized and presented [as] that of the divinatory and poetical historical saga. Divination means the vision of the historical emergence which precedes 'historical' events and which can be guessed from that which has emerged and in which 'historical' history takes place. And poetry means the articulated form of this divining vision and therefore of the historical emergence seen in this way. In this kind of divination and poetry narrative saga arises distinct from history and in connexion with it

– and woe to the history which lacks this connexion" (pp. 82–83). In myth, "both the gods and the story are not the real point at issue, but only point to it. The real object and content of myth are the essential principles of the general realities and relationships of the natural and spiritual cosmos which, in distinction from concrete history, are not confined to definite times and places" (p. 84).

[41] Ibid., p. 87ff.

[42] Ibid., pp. 243ff.

[43] *CD*, III/1, p. 340.

[44] Emil Brunner, *Revelation & Reason*, trans. Olive Wyon (Philadelphia: Westminster, 1946), pp. 222ff.

[45] Brunner, p. 228.

[46] Ibid., p. 231.

[47] Ibid., pp. 232, 234.

[48] Ibid., p. 236.

[49] Here Brunner includes Buddhism as well, arguing that the Buddha never claimed to have received divine revelation. Despite the enlightenment experience of Buddha, Brunner argues that it differs from revelation because no self–revealing God is involved (*Revelation & Reason*, pp. 225ff.).

[50] Paul Tillich, *Systematic Theology*, vol. III (New York: Harper & Row, 1967), p. 367.

[51] Tillich, p. 367.

[52] Paul Tillich, *The Future of Religions*, ed. by Jerald C. Brauer (New York: Harper & Row, 1966), pp. 86ff.

[53] Paul Tillich, *Christianity and the Encounter of the World Religions* (New York: Columbia University Press, 1963), pp. 53ff.

[54] Tillich, *Christianity and the Encounter of the World Religions*, p. 97. It must be pointed out that Tillich does not follow his own dictate – he reduces Christianity to a concept, his words about the centrality of the Christian revelation notwithstanding. By starting with the concept of religion, he has, *a priori*, begun the reduction of the religion that he places at the center of everything; but then, that may be his wish for Christianity, since its diffusion, he points out, is the only manner by which it can last.

[55] Ibid., pp. 64ff. Though to be fair, in his introduction, Tillich maintains that these Bampton Lectures were "not supposed to give embracing answers to the manifold problems raised in the discussion of the subject matter, but to confront the reader with some points of view which I consider decisive for every approach to the central problem" (p. vii).

[56] P. Tillich, *The Future of Religions*, p. 91. Cf. also Tillich's comments on theology and art in *Systematic Theology III*, pp. 196ff.

[57] Karl Rahner, *Theological Investigations*, vol. 5 (London: Darton, Longman & Todd, 1966), pp. 115ff.

[58] Paul Knitter, "Introduction", *The Myth of Christian Uniqueness – Toward a Pluralistic Theology of Religions*, p. viii. See also our mention of Kärkkäinen, Dupuis, and Knitter's new model in footnote #7.

[59] See Harold Netland, *Encountering Religious Pluralism*, pp. 48ff.

60 Jurgen Moltmann, *The Church in the Power of the Spirit* (London: SCM Press, 1977), p. 153. We must also note that Karl Barth's Christology is equally expansive in this area. See *Church Dogmatics*, IV/1, pp. 316ff.

61 See Stanley J. Samartha, "The Cross and the Rainbow: Christ in a Multireligious Culture," Raimundo Panikkar, " The Jordan, the Tiber, and the Ganges: Three Kairological Moments of Christic Self–Consciousness," Seiichi Yagi, "'I' in the Words of Jesus," and Aloysius Pieris, "The Buddha and the Christ: Mediators of Liberation." All in *The Myth of Christian Uniqueness*, ed. Hick & Knitter.

62 Gerald McDermott is one of the few who takes the discussion towards revelation and away from salvation in his *Can Evangelicals Learn from World Religions?* (2000)

63 McDermott takes a small step with three chapters about Buddhist no–self and no–mind, a Daoist theory of camouflage, and Mohammad and the signs of God (McDermott, pp. 133ff.)

64 John Robinson, *Truth is Two–Eyed* (Philadelphia: Fortress Press, 1979).

65 John Cobb, *Beyond Dialogue: Towards a Mutual Transformation of Christianity and Buddhism* (Philadelphia: Fortress Press, 1982), pp. 99ff.

66 Cobb, pp. 123ff. It is interesting to note however, that Cobb's suggestions about what Christianity can learn from Buddhism are far more numerous and detailed, than what Buddhism can learn from Christianity.

67 Raimundo Panikkar, *The Unknown Christ of Hinduism*, rev. ed. (Maryknoll, NY: Orbis, 1979), p. 29.

68 John Hick, *Death and Eternal Life* (New York: Harper & Row, 1976), p. 27.

69 Hick, *Death and Eternal Life*, pp. 30–33.

70 Ibid., pp. 413–414.

71 Ibid., p. 416.

72 Ibid., p. 416. Note: As one might expect in Hick's pluralism he has dropped Christianity's capital "C".

73 Cf. Hick, pp. 30–33. Cf. also the aforementioned work of Gadamer, pp. 24ff.

74 Ibid., p. 418.

75 Ibid., p. 427.

76 Ibid., p. 461.

77 Ibid., p. 464.

78 Ibid., p. 428.

79 S. Mark Heim, *The Depth of the Riches: A Trinitarian Theology of Religious Ends* (Grand Rapids: Eerdmans, 2001), p. 6.

80 Heim, p. 7.

81 Ibid., p. 7.

82 Ibid., p. 44. See also Joseph DiNoia, *The Diversity of Religions: A Christian Perspective* (Washington, D.C.: Catholic University of America Press, 1992), p. 91.

83 See Heim, p. 119.

84 Heim, p. 44.

85 Ibid., p. 49.

86 Ibid., p. 9.

[87] Ibid., p. 164.

[88] Ibid., p. 76.

[89] On whether Heim's theory of religious ends is plausible or not, Jacques Dupois concludes that the theory of diverse religious ends is "deceptive and unconvincing" in *Toward a Religious Theology of Religious Pluralism*, pp. 309–313. Further, we have labeled Heim an "inclusivist," a term he might reject. However Kärkkäinen's critiques mirror our own: "This raises the question of whether Heim's proposal is pluralistic at all. I doubt it… In his theology of religions, Heim takes for granted the idea of the Christian God and does not even question its relevance for the rest of religions and their religious ends, an inevitable conclusion of his interpretation being, for example, that the validity of *nirvana* for the Buddhist end is based on the triune God (see 2001, especially p. 179). But this claim is, of course, nothing other than a typical inclusivist reading of other religions, totally against their own self–understanding, thus making Heim's critique of the faults of other kinds of pluralisms begin to falter." See Kärkkäinen, *Trinity and Religious Pluralism*, p. 149.

[90] Smart & Konstantine, p. 441.

[91] E.g. "There is, first, the infinity of the divine life as it circulates through the selfless spirits. This is the non–relational aspect. Then, second, there is the plurality of the three Persons. Third, there is the communal life – the shared ego of the three. These last two aspects are relational (the first to one another, the second toward creatures)" (Smart & Konstantine, p. 174).

[92] Ibid., pp. 174ff.

[93] Ibid., p. 297.

[94] We must also note, however, that even in the case of serious exclusivists, there has remained a strong pastoral concern for those of other traditions. Augustine and Aquinas certainly cared for those who believed differently, even if they felt others were misdirected or infidels. So the caricature of angry, militant Christians, apoplectic about the presence of other religions is simply not universally applicable in the history of the church. One has been able to hold both to the exclusivisity of Christ's salvation and to respect and concern for other views in the Christian tradition.

Chapter Five

[1] Wisdom literature and the doctrine of creation are also similar in that they both have tended to be subordinated to other more dominant concerns in theology, e.g. redemption (see Leo G. Perdue, *Wisdom & Creation* (Nashville: Abingdon, 1994, p. 19).

[2] Preuss takes the basic view that early wisdom (e.g. Prov. 10ff.) is an essentially foreign religion within the OT. International in character, and focusing almost exclusively on human activity within a deed/consequence framework, wisdom in this context for Preuss is a religion different from the rest of the OT conception. See H.D. Preuss, *Einfürung in die alttestamentlich Weisheitsliteratur* (Urban – Taschenbumlautcher 383; Stuttgart: Kohlhammer, 1987). For how Preuss builds on the earlier deed/consequence work of Koch, see Roland E. Murphy, *The Tree of Life*, 2nd ed. (Grand Rapids: Eerdmans, 1996), p. 129, n. 21.

[3] Here we are following the work of both H.H. Schmid, "Creation, Righteousness and Salvation: 'Creation Theology' as the Broad Horizon of Biblical Theology", in B.W. Anderson [ed.], *Creation in the Old Testament* [London, 1984], pp. 102–117; and W. Zimmerli, in "The Place and Limit of the Wisdom in the Framework of the Old Testament Theology", *Scottish Journal of Theology*, 17 (1964), pp. 146–158. This is reprinted in James L. Crenshaw's fine anthology, *Studies in Ancient Israelite Wisdom* (New York, Ktav, 1976), pp. 314–326.

[4] Zimmerli, p. 148.

[5] Ibid., p. 152. Here Zimmerli does not emphasize any environmental implications of this verse.

[6] In another article, Zimmerli also accentuates the anthropological thrust of wisdom. W. Zimmerli, "Concerning the Structure of Old Testament Wisdom", trans. Brian W. Kovacs, reprinted in J.L. Crenshaw's *Studies in Ancient Israelite Wisdom*, pp. 175 – 205.

[7] Zimmerli, "The Place and Limit of the Wisdom in the Framework of the Old Testament Theology", p. 153.

[8] Ibid., pp. 156–157. Von Rad writes, "Ecclesiastes is anything but a nihilistic agnostic. He knows that the world was created by God and that he unceasingly acts upon it" (Von Rad, *Old Testament Theology*, vol. 1, p. 456).

[9] J.L. Crenshaw, "Prolegomenon", in *Studies in Ancient Israelite Wisdom*, p. 28.

[10] The issue of how or whether Israel followed a rather strict theory of retribution is a hotly debated one in OT theology. See G. Von Rad, *Wisdom in Israel* (Nashville: Abingdon, 1972), who articulates how for the Sages the freedom of God transcended any retributive system.

[11] Crenshaw, p. 34. We don't want to overemphasize Crenshaw's position at this point, however. Many have argued for the primacy of the family in wisdom's development (especially in Proverbs), and not a general malaise in society. Murphy notes, "the home may be regarded as perhaps the original site of wisdom teaching, before and after such teaching became professionalized among the sages" (Murphy, *Tree of Life*, p. 4). The role of the clan is critically important to any discussion of wisdom's background.

[12] Ernest Würthwein, "Egyptian Wisdom and the Old Testament", *Studies in Ancient Israelite Wisdom*, p. 119.

[13] See Roland Murphy's examination of the role of *ma'at* and the personification of wisdom in Proverbs 1–9 in *Tree of Life*, pp. 161ff., and our treatment (pp. 123ff.). Another significant subtheme to be mentioned in introducing wisdom literature is that of the anthropological nature of wisdom. Crenshaw notes that wisdom is a "way of looking at things [that] begins with humans as the fundamental point of orientation. It asks what's good for men and women, and it believes that all essential answers can be learned in experience, pregnant with signs about reality itself" (James L. Crenshaw, *Old Testament Wisdom* [Louisville: Westminster John Knox, 1998], p. 10). Wisdom does not begin with an overtly theological thrust, but with a more humanistic perspective – e.g. "what is required to live right?" We have already noted that this perspective arose in a time of great social and religious upheaval, which has led some scholars to emphasize that wisdom, in general, is out of character with the rest of Old Testament theology, with its heavy emphasis on salvation history. In Walther Zimmerli's article on the structure of wisdom, he points out that even though wisdom's autonomous man wants to organize himself freely outwards and assess the world, in the end by attempting to judge himself

by an order binding man, he succumbs to a position that is not radically anthropocentric (W. Zimmerli, "Concerning the Structure of Old Testament Wisdom", p. 177). However, this does not mean that the theological element is stressed – e.g. Proverbs, Job, and Ecclesiastes make no reference to humanity as a member of the covenant people. Furthermore, "God is seen from man's viewpoint, and his conduct … is evaluated against the worth that it can have for man" (Zimmerli, p. 187). Though the Creator's authority is referenced, the role and expectations of humanity are ultimately the prevailing subjects (Zimmerli's perspective in this article is quite modern and romantic, almost Nietzchean–sounding, as he speaks of "unbound and self–determining man" scrutinizing his life and world [p. 189]. In his later article ["The Place and limit of the Wisdom in the Framework of the Old Testament Theology"], his analysis of wisdom is much more theological).

While it is clear that wisdom has an anthropological thrust, the pure secularism that some claim for wisdom does not follow; others argue for wisdom assuming a divine order. For instance, Crenshaw argues that both H. J. Hermisson and Schmid challenge this solely anthropological viewpoint. Hermisson writes that man is measured against the world in which he is placed, not in solitary measure against himself. Schmid argues that cosmology and cosmic order is the center, not man (see Crenshaw, "Prolegomenon", p. 43, n. 99). Boström, following Zimmerli, posits that Israelite wisdom tradition relies too heavily on the reality and interrelationships of a divine order for secularism to be an appropriate term (L. Boström, *The God of the Sages* [Stockholm, 1990], pp. 37, 136). And in refuting the claim that there was a secular "older wisdom" that was augmented by a more theological wisdom in Israel, Crenshaw posits that the development of wisdom is not from secularism to theological reflection, but rather 1) family wisdom; to 2) court wisdom; to 3) theological wisdom. All of these contained a religious element from the beginning (See Crenshaw, *Old Testament Wisdom*, pp. 77ff.).

14 Crenshaw, p. 44.

15 The place–name 'Massa', not the word for 'oracle', is in view here. See John Day, "Foreign Semitic influence on the wisdom of Israel and its appropriation in the book of Proverbs", in *Wisdom in Ancient Israel*, ed. John Day, Robert P. Gordon, and H.G.M. Williamson (Cambridge, 1995), p. 55.

16 Crenshaw, *Old Testament Wisdom*, p. 206.

17 G. Buccellati questions whether there is a wisdom literature per se in Mesopotamia. He is willing to accede wisdom "themes" that occur in various types of literature, but not a literary genre. He states, "Wisdom should be viewed as an intellectual phenomenon in itself. It is the second degree reflective function as it begins to emerge in human culture." In "Wisdom and Not: The Case of Mesopotamia", *Journal of the American Oriental Society,* 101/1 (1981), p. 44. While Buccellati may very well be correct, it does not preclude us from accepting W.G. Lambert's observation that "wisdom" has been used "for a group of texts which correspond in subject–matter with the Hebrew Wisdom books, and may be retained as a convenient short description." See *Babylonian Wisdom Literature* (Oxford: Clarendon, 1960), p. 1.

18 It may be that there is an Edomite background for the book of Job. See John Day, "Foreign Semitic influence on the wisdom of Israel and its appropriation in the book of Proverbs", *Wisdom in Ancient Israel*, pp. 56–57.

[19] Day, p. 57. Day also suggests "possibly it was the ancient Egyptians' strong belief in immortality, combined with their traditional fatalism, that prevented the problem of theodicy from raising its head in the way in which it did in the Semitic world" (p. 57).

[20] W.G. Lambert, "Some new Babylonian wisdom literature", *Wisdom in Ancient Israel*, Ed. J. Day, p. 37.

[21] Lambert, "Some new Babylonian wisdom literature", p. 41.

[22] See R. Murphy, p. 159.

[23] Crenshaw, *Old Testament Wisdom*, p. 223.

[24] Würthwein, "Egyptian Wisdom and the Old Testament", p. 114.

[25] James B. Pritchard, ed., *The Ancient Near East*, vol. 1 (Princeton: Princeton University Press, 1958), p. 243. Cited hereafter as *ANE*.

[26] Pritchard, *ANE*, p. 238.

[27] Glendon E. Bryce, *A Legacy of Wisdom* (New Jersey: Associated University Presses, 1979), pp. 58ff.

[28] Bryce, p. 104ff.

[29] Ibid., pp. 193ff.

[30] See Crenshaw's discussion of Niccacci, Römheld, and Whybray for further analysis of Semitic or Egyptian dependence in *Old Testament Wisdom*, p. 83, n. 12. In the end, however, it may not be completely crucial to establish which source came first; it is clear that there is at the very least a mutual *interdependence*.

[31] See J.D. Ray, "Egyptian Wisdom Literature", in *Wisdom in Ancient Israel*, Ed. J. Day, pp. 23–24.

[1] See Day for the previous chart as well, pp. 59–60.

[2] Cf. R.N. Whybray, *Ecclesiastes* (Grand Rapids and London, 1989), pp. 143–144.

[3] See Bruce W. Jones, "From Gilgamesh to Qoheleth", in *The Bible in the Light of Cuneiform Literature*, ed. W.W. Hallo, B.W. Jones and G.L. Mattingley (Lewiston: Edwin Mellon Press, 1990), pp. 366–367.

[4] Jones, p. 397.

[5] See Jones, pp. 370–373. Jones admits, however, the evidence is too limited to be completely certain.

[6] Day, p. 59.

[7] Crenshaw, *Old Testament Wisdom*, p. 123.

[8] Ibid., p. 124. Perhaps this is because Qoheleth knows Genesis.

[9] Würthwein, pp. 123ff.

[10] Würthwein goes on to note that in Ecclesiastes, in contrast to the limitedness of humanity, we see the utter freedom of God: "Here, God appears entirely sovereign, bound to no consideration of human endeavor or a moral order. Qoheleth, however, does not feel himself driven to protest against God, whose caprice, or better enigmatic character, he has recognized, as Job did. He only confirms it. This confirmation, though, means a total denial of the image of God in wisdom" (p. 126). Würthwein goes on to offer that humanity's response to this freedom is then fear; not reverence or piety, but rather something more original, a sense of mysterious awe before an overwhelming force.

[11] Here we may also note that Gilgamesh is an epic, and as such is not as concerned with making such broad theological and anthropological claims as Qoheleth is. The similarities and differences between the two happen at intersections, where the story of Gilgamesh meets with the teachings of a sage.

[12] Cf. the Mesopotamian *Dialogue of Pessimism*, the Egyptian *Papyrus Insinger* and *The Lamentations of Khakheperre–sonbe*.

[13] *ANE*, pp. 148–160.

[14] W.S. Lasor, D. Hubbard, F.W. Bush, *Old Testament Survey* (Grand Rapids, Eerdmans, 1982), p. 540. H. Gese sees in *I Will Praise the Lord of Wisdom* the literary form of the "paradigm of an answered complaint". Cf. *Lehre und Wirklichkeit in der alten Weisheit* (Tübingen: Mohr/Siebeck, 1958).

[15] Lambert, *Babylonian Wisdom Literature*, p. 27.

[16] Lambert, "Some New Babylonian Wisdom Literature," pp. 32ff.

[17] Ibid., p. 37. Other ancient near eastern works with affinities to Job include the *Egyptian Dispute over Suicide (Man Who Was Tired of Life)*, and the *Tale of the Eloquent Peasant*.

[18] Murphy, p. 133.

[19] See the general survey in C.V. Camp, *Wisdom and the Feminine in the Book of Proverbs* (Sheffield, 1985), pp. 23–46.

[20] C. Kayatz, *Studien zu Proverbian 1–9* (Neukirchen–Vluyn, 1966), pp. 76–119.

[21] Von Rad, *Wisdom in Israel*, p. 153.

[22] Day, p. 68.

[23] See H. Ringgren, *Word and Wisdom* (Lund, 1947), p. 144.

[24] M.S. Smith, *The Early History of God* (San Francisco, 1990), pp. 94–95.

[25] Day, p. 69.

[26] Ibid., p. 70.

[27] Judith M. Hadley, "Wisdom and the goddess", in *Wisdom in Ancient Israel*, Ed. J. Day, p. 236.

[28] See Judith M. Hadley's *Evidence for a Hebrew Goddess: The Cult of Asherah in Ancient Israel and Judah* (Cambridge: Cambridge University Press, 2000).

[29] See Murphy, p. 136.

[30] Here we may also note that this enjoyment of creation is centered theologically for Moltmann in the Spirit (cf. *God in Creation*).

[31] See Roland E. Murphy, "The Personification of Wisdom", in *Wisdom in Ancient Israel*, Ed. J. Day, pp. 225–226.

[32] This world order is centered around "personified wisdom", which he calls "the self–revelation of creation" (cf. ch. 9 of *Wisdom in Israel*). Von Rad still wants to maintain a distinction between God and creation with Lady Wisdom: "And yet, wisdom is not Jahweh himself; it is something separate from him: indeed it once designates itself as Jahweh's creature, albeit the first born of all creatures (Prov. 8:22), and identifies itself with the thoughts which God cherished in creating the world (Prov. 3:19)." (von Rad, *Old Testament Theology*, vol. I, p. 444). Murphy counters this by noting, "the very origins and the authority of Wisdom suggest more than a personified order of creation. Wisdom is somehow identified with the Lord. The call of Lady Wisdom is the voice of the Lord; she is the revelation of God, not merely the self–revelation of creation. She is

the divine summons issued in and through creation, sounding through the vast realm of the created world, and heard on the level of human experience" (Murphy, *The Tree of Life*, p. 138).

[33] von Rad, *Wisdom in Israel*, p. 156.

[34] von Rad, *Old Testament Theology*, vol. I, p. 444.

[35] Though here Murphy notes that Lady Wisdom is quite a broad and multi–faceted figure: "The functions of Lady Wisdom are manifold; they are as broad as life itself, in keeping with her eternal origins from the source of all life.... this open–ended character of Lady Wisdom enabled future sages to make further additions and thus achieve a formidable description of her self and her activities" in *The Tree of Life*, p. 137.

[36] Murphy, "The personification of wisdom", p. 232.

Chapter Six

[1] I use the term journey consciously, as opposed to something such as "building block" imagery, which for me, is too foundationalist. Rather than establishing my argument on "indubitable principles", it is better to use travel imagery – i.e. after making necessary stops in one area, we are then free to move on to the next area. This I believe is more true to a narrative approach and a holistic epistemology that is key for postmodernity.

[2] Arvind Sharma makes this point clearly in attempting to define Hinduism. Vedic authority may be the sole test of orthodoxy in Hinduism, yet this authority is quite tenuous. For Sharma, being a Hindu is much more about a way of life than adherence to a text (see "Hinduism", *Our Religions,* ed. Arvind Sharma {New York: Harper Collins, 1993}, p. 5).

[3] W.C. Smith, "A Human View of Truth" in *Truth and Dialogue*, ed. J. Hick (London: Sheldon, 1974), p. 20.

[4] See Charles H. Long's article "Cosmogony" in *The Encyclopedia of Religion*, vol. IV, ed. Mircea Eliade (New York: Macmillan, 1987), pp. 94–100, for the various options.

[5] Kees W. Bolle includes a helpful delineation between cosmology and worldview: "*Worldview* is the term for a more general, less precisely delineated, but commonly accepted set of ideas (an "ideology") concerning life and the world. *Cosmology* refers to more consciously entertained images, doctrines, and scientific views concerning the universe." In "Cosmology", *The Encyclopedia of Religion*, vol. IV, ed. Mircea Eliade (New York: Macmillan, 1987), p. 100.

[6] Maseo Abe, "Buddhism", *Our Religions*, p. 72.

[7] Abe, "Buddhism", p. 75.

[8] Ibid., p. 115.

[9] Ibid., pp. 73–74.

[10] W. Randolph Kloetzli, *Buddhist Cosmology: Science and Theology in the Images of Motion and Light* (Delhi: Motalal Banarsidass, 1989), p. 12.

[11] Kloetzli, *Buddhist Cosmology*, p. 3.

[12] The third cosmology focuses on the ten "fields" of Buddha, and the powers of the Buddha displayed therein. These powers are repeatedly described as being immense, incalculable, and inconceivable. Here, the sense of the miraculous and spectacular is

overarching. See, for example, the rebirth in *Sukhāvatī*. Cf. Allan A. Andrews, *The Teachings Essential for Rebirth* (Tokyo: Sophia University, 1973).

[13] Vasubandhu, "*Abihiddarmakosbhasyam*", (vol. II) rendered by Louis de La Vallée Poussin, Eng. Trans. By Leo M. Pruden (Berkeley, CA: Asian Humanities Press, 1988), pp. 451–452.

[14] *The Mahavastu*, trans. J. J. Jones (London : Luzac, 1949), vol. 1, pp. 96–98. De La Vallēe Poussin writes that the "chilicosm did not satisfy the Buddhist imagination." L. De La Vallēe Poussin, "Cosmology and Cosmogony (Buddhist)", *Encyclopedia of Religion and Ethics*, ed. James Hastings, vol. IV, (Edinburgh: T & T Clark, 1911), p. 122. Of interest here is whether in the absence of a deity, the Buddhist imagination has substituted the concept of an infinite universe for that of an infinite God. Or alternately, whether Buddhism has retained a construct for transcendence.

[15] Kloetzli, "Buddhist Cosmology," *The Encyclopedia of Religion, vol. IV*, ed. Mircea Eliade (New York: Macmillan, 1987), p. 117.

[16] Nāgārjuna, *Madhyamakakārikās*, (1:1) in Frederick J. Streng, *Emptiness* (Nashville: Abingdon, 1967), p. 213.

[17] Kloetzli, *Buddhist Cosmology*, p. 2.

[18] De La Vallēe Poussin, p. 130. Paul Martinson writes that Buddhism has no doctrine of creation in, "Wisdom and Love as the Basis for Preaching in Buddhism and Christianity," *Christianity and the Religions of the East*, ed. Richard W. Rousseau (Scranton, PA: Ridge Row Press, 1982), p. 120.

[19] Winston L. King, *A Thousand Lives Away: Buddhism in Contemporary Burma* (Cambridge, MA: Harvard University Press, 1964), pp. 108–109.

[20] Leo D. Lefebure, *The Buddha & the Christ* (Maryknoll, NY: Orbis, 1993), p. 47.

[21] Abe, p. 115.

[22] Daisetz Teitero Suzuki, *Studies in the Lankāvatāra Sūtra* (London: Routledge & Kegan Paul Ltd., 1930), pp. 148–153.

[23] Lefebure, p. xx.

[24] Khin Maung Din, "Some Problems and Possiblities for Burmese Christian Theology Today," *Christianity and the Religions of the East*, ed. Richard W. Rousseau (Scranton, PA: Ridge Row Press, 1982), p. 81.

[25] Din, p. 80.

[26] Ibid., p. 81.

[27] One thinks here of Paul Tillich's comments on personalism in *Biblical Religion and the Search for Ultimate Reality* (Chicago: University of Chicago Press, 1955), pp. 21ff.

[28] John B. Cobb, Jr. & Christopher Ives, ed. *The Emptying God: A Buddhist–Jewish–Christian Conversation* (Maryknoll, NY: Orbis, 1990).

[29] Moltmann, *God in Creation*, p. 56.

[30] Walpola Sri Rahula, *What the Buddha Taught* (New York: Grove Press, 1974), p. 51.

[31] An interesting correlation is seen here between Buddhist thinking and some radical postmodernists such as Mark C. Taylor, who posits that the self, or soul, in postmodernity, has been decentered, and ultimately, is dead (in contrast to the earlier Romantic, Modern conception). See Mark C. Taylor, *Erring* (Chicago: University of Chicago Press, 1984), p. 141.

[32] See Ray S. Anderson's discussion in *On Being Human*, pp. 215ff.

[33] Abe, p. 119.

[34] Paul Knitter, "Horizons on Christianity's New Dialogue With Buddhism", *Christianity and the Religions of the East*, ed. Richard W. Rousseau (Scranton, PA: Ridge Row Press, 1982), p. 95.

[35] See below in section on Hindu cosmogony (pp. 178ff.).

[36] Knitter, p. 95.

[37] Here Barth's stress of a non–anthropocentric starting point for theology is crucial, though Barth did not emphasize the immanent implications in his Trinitarian theology, especially in the work of the Holy Spirit.

[38] Ninian Smart, *Buddhism and Christianity: Rivals and Allies* (Honolulu: University of Hawaii Press, 1993), pp. 81–82.

[39] Arvind Sharma, "Hinduism", *Our Religions*, p. 5. Further, Roger Hooker argues that the word "Hinduism" is a Western invention, aimed at distilling (imposing?) a phenomenon into an organized system of beliefs and practices. See Roger Hooker, *Themes in Hinduism and Christianity* (Frankfurt: Verlag Peter Lang, 1989), p. 5.

[40] Sharma, "Hinduism," *Our Religions*, p. 7.

[41] Ibid., p. 24. Cf. also Arvind Sharma, "The Puruṣāthas: An Axiological Exploration of Hinduism", *Journal of Religious Ethics*, 27:2.

[42] Ibid., p. 36.

[43] See J.L. Brockington, *Hinduism and Christianity* (New York: St. Martin's Press, 1992), p. xi. Brockington argues that it is possible to be an atheist and still be a Hindu, because practice, not belief is paramount.

[44] H. Jacobi, "Cosmogony & Cosmology (Indian)", *Encyclopedia of Religion and Ethics*, ed. James Hastings, vol. IV, (Edinburgh: T & T Clark, 1911), p. 156.

[45] Cf. R.C. Zaehner, *Hindu Scriptures* (London: Dent, 1966), pp. 11–12.

[46] Barbara A. Holdrege, *Veda and Torah: Transcending the Textuality of Scripture* (New York: State University of New York Press, 1996), p. 43.

[47] *Śatapatha Brāhmana*, XI.1.8.3. *The Śatapatha Brāhmana*, according to the text of the *Madhyandina* School, trans. by Julius Eggeling (Oxford: The Clarendon Press, 1889–1900).

[48] *Śatapatha Brāhmana*, VI.1.1.8–10.

[49] See Holdrege, *Veda and Tora*, pp. 52ff.

[50] R.C. Zaehner, *Hindu Scriptures*, pp. 8–9.

[51] W. Randolph Kloetzli, "Hindu and Jain Cosmologies," *The Encyclopedia of Religion*, vol. IV, ed. Mircea Eliade (New York: Macmillan, 1987), p. 108. Cf. also Gerald J. Larson, *Classical Sāṃkyha: An Interpretation of Its History and Meaning*, 2nd ed. (Santa Barbara, 1979).

[52] Holdrege, p. 65.

[53] *Bṛhad–āraṇyaka Upaniṣad*, V.5.1. Cf. *The Principal Upaniṣads*, ed. S. Radhakrishnan (New York: Harper & Brothers, 1953.

[54] Kloetzli, "Hindu and Jain Cosmologies", pp. 107–108.

[55] Ibid., p. 109.

[56] Jacobi, p. 158ff. Cf. H. H. Wilson, trans. and ed. *The Vishnu Purāṇa* (1840), 2nd ed. (Calcutta, 1961).

[57]

[58] See Brockington, p. 7. Cf. Sarvepalli Radhakrishnan and Charles Moore, *A Sourcebook in Indian Philosophy* (Princeton, 1957), pp. 37–96.

[59] Brockington, p. 8.

[60] Surendra Nath Dasgupta, *A History of Indian Philosophy*, vol. III (Cambridge: Cambridge University Press, 1952), pp. 156ff.

[61] Brockington, p. 12.

[62] Cf. Dasgupta, *A History of Indian Philosophy*, pp. 165ff.

[63] Eberhard Jüngel, *The Doctrine of the Trinity: God's Being is in Becoming* (Grand Rapids: Eerdmans, 1976), pp. xi – xviii.

[64] Jüngel, pp. 22ff.

[65] Sharma, p. 30.

[66] Brockington, p. 21. Brockington also explores how these three deities came to be grouped in such a way.

[67] See R. Panikkar, *Trinity in World Religions* (CLS, Madras: 1970).

[68] Frank Whaling, "The Trinity and the Structure of Religious Life", *Christianity and the Religions of the East*, ed. Richard W. Rousseau (Scranton, PA: Ridge Row Press, 1982), p. 45. Whaling cites Keshab Chandra Sen, a 19th Century Indian theologian, who suggested that the Trinity and *Saccidānanda* were similar notions of the Godhead. "The Father … is the Creator, the still God, the "I am" of the Godhead, who is Force, the True, and *Sat*. The Son … is the Example, the journeying God, the "I love" of the Godhead, who is Wisdom, the Good, and *Cit*. The Holy spirit … is the sanctifier, the returning god, the "I save of the Godhead, who is Holiness, the Beautiful, and *Ānanda*." (p. 46).

[69] Whaling, pp. 46–47.

[70] See Brockington, p. 134.

[71] E.g. Lynn White's critiques of the religious roots of the environmental crisis in "The Historical Roots of our Ecological Crisis," *Science* 155 (March 10, 1967). We will comment further on White's position in our last chapter.

[72] Cf. Seyyed Hossein Nasr, "Islam", *Our Religions*, ed. Arvind Sharma (New York: Harper Collins, 1993), p. 456.

[73] We are using Ali's contemporary translation. Ahmed Ali, *Al–Qur'ān* (Princeton: Princeton Unversity Press, 1988).

[74] Though Allah is *the* dominant figure, Nasr and others are quick to point out the "plurality" of Islam, arguing that Islam begins with the first prophet, Adam, who is representative of a line of prophets stretching from Moses to Christ, and ending with Mohammed. The Qur'ān states this: "Mankind were one community, and Allah sent (unto them) Prophets as bearers of good tidings and as warners …" (2, 213). It was primarily the school of Sufism and Islamic metaphysicians who opened the door to a more pluralistic understanding of Islam (see Nasr, p. 522). However, this sense of plurality still seems to be confined to primarily an Islamic/Christian/Jewish nexus, and not one encompassing other Eastern religions. Furthermore, the plurality seems to be based on an anthropological understanding, and not a theological standing. There is one humanity,

Islam seems to argue, and out of this flows an openness to other religions. However, from the basis of one God, Islam falters at being an "incorporating" religion.

[75] Nasr, p. 450. This perspective is Sufi thought, which has more of an emphasis on inner piety. Interestingly, the Qur'ān reverses this order: God only loves those who love him (cf. 3, 31–32).

[76] Ibid., p. 451.

[77] For the *Hādith,* see Allama Sir Abdullah al–Ma'mun al–Suhrawardy, *The Sayings of Muhammad* (New York: Carol Publishing Group, 1990).

[78] Nasr, p. 445.

[79] Ibid., p. 464.

[80] J. Windrow Sweetman, *Islam and Christian Theology*, vol. II (London: Lutterworth Press, 1947), p. 157.

[81] Nasr, pp. 465–466. F. Schuon writes of the "pillars of Islam" being the double testimony of faith, the canonical prayer repeated five times a day, the fast of Ramadan, the tithe, the pilgimage, and to these is sometimes added the holy war. See Frithjof Schuon, *Understanding Islam*, trans. D.M. Matheson (New York: Roy, 1963), pp. 38–39.

[82] Here we must also note that this inner emphasis is greater in Shiʿism than Sunnism. The vast majority of Muslims – about 87 – 88 percent – are Sunnis, that is, followers of the *sunnah* (or reports or words) of the Prophet. About 12 to 13 percent of Muslims are Shiʿites. Shiʿites are divided into Twelve–Imām Shiʿites, Ismāʿilis, and Zaydis. The Twelve–Imām Shiʿites are by far the most numerous, comprising some 130 million people living mostly in present–day Iran, Iraq, Lebanon, the Persian Gulf, eastern Saudi Arabia, Afghanistan, Pakistan, and India (see Nasr, pp. 431–432).

[83] Schuon, p. 40.

[84] Ibid., p. 158. One consequence of this emphasis on intelligence is the charge of gnosticism leveled at Islam. Schuon discusses this in detail (see pp. 109ff.). This may apply to a doctrine of creation as well. In emphasizing the importance of a 1st principle – Logos – that may guide creation, Islam seems to have set up a Platonic form, a principle outside of God, to which ontology must conform. This is precisely the difficulty modern Christian theology had when inserting "truth" and "rationality" as arbiters of theological debate in the last century. Postmodern theology has improved upon this foundationalist epistemology by asserting a non–foundationalist epistemology, seen notably in the work of a Karl Barth, whose theology is deeply critical of 19th century Liberalism. Barth's position – that no epistemology, theory of knowledge or even scripture can account for God – sounds closer to Islam's roots than perhaps some of the doctrine surrounding Islam's doctrine of creation does.

[85] R. Arnaldez, "Khalḳ", *The Encyclopaedia of Islam*, vol. III, ed. B. Lewis, et.al. (Leiden, E.J. Brill: 1971), p. 981. I am indebted to Arnaldez for much of the following lexicographical data.

[86] Arnaldez, p. 982.

[87] Ibid., p. 982.

[88] Arnaldez, p. 982. The use of the word "viceroy" is a more contemporary translation of the word. Many scholars would say the word – "*caliph*" – originally meant "successor", as in humanity being the true "*caliph*" of Allah on earth.

[89] Ibid., p. 984. L. Gardet reads the literature differently in considering *ibdā',* or "absolute creation" or "primordial innovation". Though *ibdā* is not Qur'ānic, it is to be distinguished from *khalḳ* and *takwīn* as follows: "*Ibdā',* then, denotes the bringing into existence ... of a thing, "without anteriority of matter or of time".... *Ibdā'* is thus placed in correlative opposition to *takwīn,* defined as the production of being with anteriority of matter and time. The distinction between *ibdā'* and *khalḳ* is as follows: the former denotes bringing into existence with nothing preceding; the second, bringing into existence from an existing thing." See L. Gardet, "*Ibdā',*" *The Encyclopaedia of Islam,* vol. III, ed. B. Lewis, et.al. (Leiden, E.J. Brill: 1971), pp. 664–665. Gardet's comments highlight the sense of fashioning that is a part of the Islamic sense of creation, though in the end a doctrine of creation *ex nihilo* is paramount.

[90] Arnaldez, p. 985.

[91] For further information on the schools of Islamic theology (*kalām*), see Nasr, pp. 503ff.

[92] This implication is picked up in Neoplatonic analyses of creation in *falsafa* (Islamic philosophy), which tends more strongly toward pantheism and emanation as modes of creation. See Arnaldez' extensive analysis. See also Origen's response to this issue in Sweetman, p. 26f.

[93] Arnaldez, p. 984.

[94] See J. Moltmann, *God In Creation*, pp. 283ff.

[95] See Sweetman, pp. 42–43. He suggests that total transcendence is inimical to religion (p. 70).

[96] Ibid., p. 48.

[97] See Sweetman, p. 26.

[98] W. Montgomery Watt, *Islam and Christianity Today* (London: Routledge & Kegan Paul, 1983), p. 92. See also Schuon, p. 21.

[99] Through a Trinitarian sense of Creation (i.e. by emphasizing the role of all three members of the Trinity) we can avoid the sharp dichotomies (i.e. creation & providence) that plague a Calvinist–type system. For example, in highlighting the role of the Spirit in continual creation, Moltmann suggests, "the Spirit is the principle of creativity on all levels of matter and life" (*God in Creation*, p. 100).

[100] Cf. Sweetman, pp. 157ff. Kenneth Cragg writes, "The most characteristic descriptive of the human status before God is *'abd*, 'servant' or 'slave,' a term that is so frequent an element in Muslim names. Man the creature stands under the Divine authority in all realms. He masters the natural world, in so far as he is allowed to, only by stooping to obey the conditions it imposes. The span and course of his days are ordered by God. Man is the creature of the Divine purpose and responds to it in worship and in submission. He is a creature also in that, when he falls short of that law, the offense is seen as a lapse rather than as defiance." In *The Call of the Minaret* (New York: Oxford University Press, 1956), pp. 45–46.

[101] Nasr, p. 444. Cf. *Sahih al–Bukhari: Arabic–English*, trans. M. Muhsin Khan (Beirut: Dar Al Arabic, 1985), VIII, 160 (Bk. 14, chap. 1, trad. 246).

[102] Ibid., p. 460.

[103] See J. Dudley Woodbury, "Different Diagnoses of the Human Condition", *Muslims and Christians on the Emmaus Road*, ed. J. Dudley Woodbury (Pasadena: Zwemer Institute of Christian Studies, 1989).

[104] Nasr, p. 460.

[105] R. Arnaldez, "Insān", *The Encyclopaedia of Islam*, vol. IV, ed. E. van Donzel, et al. (Leiden : E.J. Brill, 1978), p. 1237.

[106] One area that we must mention is that of the concept of the "perfect man" (*insān al–kāmil*). Though the idea is not prevalent in the Qur'ān, it gains momentum in the commentaries and theologies, particularly as Mohammed comes to be this Perfect Man. Mohammed comes to embody the principle of that which is closest to God – Intellect – and so corresponds to the totality of the Real (cf. the "Mohammadan Reality" in Ibn 'Arabi). This Reality comes to have functions of mediation, rather than that of messenger. See R. Arnaldez' article on "*insān al–kāmi"l* in *The Encyclopaedia of Islam*, vol. IV, ed. E. van Donzel, et al. (Leiden: E.J. Brill, 1978), pp. 1239–1241.

[107] Though the Mu´tazilīs argued for the free–will of humans, but in doing so, were considered suspect by the "orthodox" Ash´arites and Hanbahtez.

[108] See E.K. Mitchell, "Cosmology and Cosmogony (Christian)", *Encyclopedia of Religion and Ethics*, ed. James Hastings, vol. IV, (Edinburgh: T & T Clark, 1911), p 143.

[109] Though this ultimate end must be seen against the practical backdrop of Buddhism, which encourages moral excellence in this life.

[110] Cf. R. Panikkar, *The Intra–Religious Dialogue* (Bangladore: Asian Trading Corporation, 1978).

[111] Here I am emphasizing the type of theological/doctrinal dialogue that is, in my view, most helpful. Others have made the very important point that it is ultimately the practical, the ethical, the life–experience that is or should be at the core of religious dialogue. Cf. W.C. Smith, "A Human View of Truth" in *Truth and Dialogue*, and Marjorie Hewett Suchocki, "In Search of Justice", *The Myth of Christian Uniqueness*, ed. J. Hick & Paul F. Knitter (Maryknoll, NY: Orbis, 1987), pp. 149ff.

[112] Even John Hick is very clear on the ultimate outcome of such a task, and is a little at odds with his pluralistic paradigm: "I'm sure that the official belief–system of each tradition is capable of desirable development and modification at many points; but this can only properly be done from within those tradition and by their own thinkers. I, as a Christian theologian, have a responsibility to try to contribute to the continuous reforming of Christian doctrine, particularly in the light of our new awareness of the other world faiths. But it is not for me to presume to tell my Muslim or Jewish or Hindu or Buddhist colleagues how to try to develop their own traditions in response to the same new awareness." John Hick, *A Christian Theology of Religions*, p. 121.

Chapter Seven

[1] Consequently, Abe remarks, "Above all, Buddhist thought emphasizes the interrelatedness of everything and thus provides a philosophical grounding for dialogue. It not only prepares us in this manner philosophically to engage in dialogue, but it also prepares us practically to face the consequences of dialogue. The dialogical interaction of religions invariably, if imperceptibly, involves change... It is not an accident that the two religions which have been most effectively engaging in dialogue in recent times are Buddhism and Christianity – two religions with prolonged experience of functioning in religiously plural environments" (Abe, *Our Religions*, p. 131).

[2] Klostermaier, Klaus K., *A Survey of Hinduism* (Albany, NY: State University of New York Press, 1994), p. 241.

[3] Klostermaier, p. 246.

[4] *Vedārthasamgraha of Rāmānuja*, ed. and trans. S. S. Raghavachar (Mysore: Ramakrishna Ashrama, 1956), 126.

[5] Cyril Glasse, *The Concise Encyclopedia of Islam* (San Francisco, Harper & Row: 1989), p. 219.

[6] Calvin, *Institutes*, vol. I, xvi, 1.

[7] Johannes Wolleb, *Compendium theologiae Christianae* (1626), in *Reformed Dogmatics*, ed. J. W. Beardslee III (New York: Oxford University Press, 1965), p. 59.

[8] Otto Weber, *Foundations of Dogmatics*, vol. I, p. 503.

[9] Weber, p. 504. Weber adds a warning here: "We are in the particular danger today of cladding *our* 'Today' with a dignity which it certainly does not deserve from the perspective of God the creator. Whereas a few decades ago it was necessary to oppose the displacement of the Creator into the distant past in the constant style of Deism, today it is necessary to reject our coercing God into our contemporaneity which only results in making God into a concept which is at our disposal, and which may be used to illuminate the puzzles of our existence" (p. 504).

[10] G.C. Berkouwer, *The Providence of God* (Grand Rapids, Eerdmans: 1952), p. 57.

[11] H. Bavinck, *Gereformeerde Dogmatick*, II, p. 569. Quoted in Berkouwer, p. 137.

[12] We will also not be pursuing the distinction between "general providence" and "special providence" (cf. Calvin).

[13] K. Barth, *CD*, III, 3, p. 26.

[14] Barth, *Learning Jesus Christ Through the Heidelberg Catechism*, trans. Shirley C. Guthrie, Jr. (Grand Rapids: Eerdmans, 1964), p. 17.

[15] Perhaps we are being wistful! Barth's strong disavowal of any form of natural theology may preclude any positive appreciation of this method. See Kärkkäinen's discussion of whether Barth is a pluralist in *Trinity and Religious Pluralism*, pp. 22ff.

[16] Of course Barth brings this around to Christ – "Providence is nothing other than God's free grace, and God's free grace in Christ is Providence." (Barth, *Learning Jesus Christ Through the Heidelberg Catechism*, p. 62). This perspective may not give creation, or the Spirit, their rightful emphasis.

[17] We do not, here in emphasizing the conceptual, want to turn this exercise into some sort of Platonic exchange of ideas. We are aware of the "flesh and blood" narrative of Scripture, of the revealing God in history, etc. The doctrines from Buddhism, Hinduism, and Islam have their own history, and are rooted in their own stories and narratives. So that while we pursue the honing of the doctrine of creation, we are cognizant to never remove the doctrines from their realistic *sitz im leben.*

[18] Martin Buber, *I and Thou*, trans. W. Kaufman (Edinburgh: T. & T. Clark, 1979), p. 62.

[19] Karl Barth, *Church Dogmatics*, III/2, p. 245.

[20] Paul K. Jewett, *God, Creation & Revelation* (Grand Rapids: Eerdmans, 1991), pp. 211–212.

[21] Stanley Grenz, *Theology for the Community of God* (Nashville: Broadman & Holman, 1994), p. 151.

[22] Grenz, pp. 231–232.

[23] Miroslav Volf, *Exclusion and Embrace* (Nashville: Abingdon, 1996), p. 30.

[24] This may have much to do with the continual social divisions present in Hindu society. However, note Heinrich Heine comments on the pantheism of the 19th century: "It is unfortunately true and we have to admit it, that pantheism has turned people into indifferentists. They have come to think that if everything is God, then one's personal concerns are surely a matter of indifference. It does not matter whether a man busies himself with clouds or antique gems, with folk songs or the anatomy of apes, with ordinary people or comic actors. But this is where the error lies: everything is not God; God is everything. God does not manifest himself to an equal degree in everything." In *The Romantic School* (New York: 1882), pp. 57f.

[25] Pannenberg, p. 32.

[26] Ibid., p. 32.

[27] Ibid., pp. 79ff. Cf. also T.F. Torrance's contribution here in *Space, Time, and Incarnation* (New York, 1979).

[28] The dangers of immanence are well–known. Moltmann and Pannenberg must walk a fine line by arguing against both dualism and pantheism, both of which impinge upon the freedom of God; the former by limiting the creator's freedom in work and in the power of his spoken word, and the latter by limiting the creator's autonomy and uniqueness unto himself. This is where the sub–doctrine of *creatio ex nihilo* becomes so important, in preserving the uniqueness of the creator.

[29] C.S. Lewis, *The Weight of Glory* (New York: Macmillan Books, 1980), pp. 18–19

[30] Ibid., p. 151.

[31] Here we must allow for the differing usage of the word "immanent" in Christian Trinitarian language. Whereas in our overall conceptual framework, immanence is contrasted over against transcendence, and so refers to the involvement of God as opposed to the distance (transcendence) of God; in trinitarian theology, immanent (following Karl Barth) refers to God as he is himself, and economic refers to what God does.

[32] Moltmann, *The Crucified God* ((New York: Harper & Row, 1974), p. 227.

[33] Moltmann, T*he Trinity and the Kingdom of God* (London: SCM Press, 1981), p. 4. Bouma–Prediger writes, "In short, Moltmann is also significant because he writes not only out of a deep recognition of the suffering Other – whether exploited peasants or the groaning earth – but out of a memory and an acknowledgement of his own suffering. Such acknowledgments are important dimensions of any theology today – in an age characterized by 'the interruption of those who suffer.'" Steven Bouma–Prediger, *Toward a New Christian Theology of Nature* (Chicago: University of Chicago Dissertation, 1992), pp. 141–142. Quote is from Rebecca Chopp, *The Praxis of Suffering: An Interpretation of Liberation and Political Theologies* (Maryknoll: Orbis, 1986), p. 2. Cf. also the writings of Bonhoeffer in this vein in *Letters and Papers from Prison* (1953).

[34] Cf. F. Schleiermacher, *The Christian Faith*, pp. 179ff.

[35] Schleiermacher, *The Christian Faith*, p. 146.

[36] See Gunton's remarks in *The Triune Creator*, p. 180.

[37] Gunton, *The Triune Creator*, p. 180. Cf. Brunner's contention: "The danger–zone has already been entered when Creation and Providence are identified with one another. For

anyone who does not admit the distinction between the creation and the preservation of the created world does not take the fact of creation seriously" (*Dogmatics*, II, 33f.).

[38] Pannenberg, *Systematic Theology*, vol. 2 (Grand Rapids: Eerdmans, 1994), p. 42.

[39] Weber states, "[God's providence] conserves created existence as something already extant and active and thus presupposes it." In *Foundations of Dogmatics*, p. 505.

[40] Pannenberg, *Systematic Theology*, vol. 2, pp. 43–44. Quoting H.L. Martensen, *Christian Dogmatics* (Edinburgh, 1898), p. 126.

[41] Klostermaier, *A Survey of Hinduism*, pp. 131–132.

[42] Ibid., p. 150.

[43] See Grenz, *Theology for the Community of God*, p. 105.

[44] Jewett, *God, Creation & Revelation*, p. 198.

[45] Pannenberg, *Systematic Theology*, I, p. 411.

[46] Tillich, *Systematic Theology*, I, p. 237.

[47] Jewett, p. 201.

[48] Barth, *CD*, I/1, p. 371.

[49] Pannenberg, *Systematic Theology*, I, p. 411–412. Pannenberg notes that Barth, too, did not want to group omnipresence with infinity, because infinity implies a break with the finite, and omnipresence could not be explained in terms of the immanence in creation that it implies (cf. Barth, *CD*, II/1, pp. 464ff.).

[50] Tillich, p. 263.

[51] Bonhoeffer, *Letters and Papers from Prison*, pp. 279ff.

[52] Gordon Kaufman, "On the Meaning of 'God': Transcendence Without Mythology," in *Transcendence*, ed. H.W. Richardson & D.R. Cutler (Boston: Beacon Press, 1969), pp. 114ff.

[53] One thinks here of the iconic Christ portrayed in the recent film *The Passion of the Christ* (2003). This Christ possesses a clear link to transcendence that is in marked contrast to earlier (modern) attempts to connect Jesus to the audience (lampooned as the "buddy Jesus" in the film *Dogma* {1999}). *The Passion of the Christ* unashamedly touts Jesus as the Messiah battling supernatural enemies to a postmodern setting. This certainly reflects director/producer Mel Gibson's catholic sensibilities, but the success of the film also speaks to postmodern humanity's willingness to be inspired by the "otherness" of transcendence.

[54] Thomas Aquinas, *Summa Theologiae*, 1a.12.13.ad 1.

[55] Though here, too, we must note the pneumatological work in revelation.

[56] Din, "Some Problems and Possibilities for Burmese Christian Theology Today", p. 80.

[57] See Clark Pinnock's analysis of the church wrongly following Greek thought in positing negative attributes of God – impassible, atemporal, etc. – in *Most Moved Mover* (Grand Rapids: Baker, 2001), pp. 71ff.

[58] Robert Jenson, *Systematic Theology*, I, (New York: Oxford University Press, 1997), pp. 211–212.

[59] Tillich, *Systematic Theology*, I, p. 245.

[60] Jewett, p. 176.

[61] Maseo Abe, "Kenotic God and Dynamic Sunyata", in *The Emptying God*, ed. John B. Cobb & Christopher Ives, p. 16.

[62] Moltmann, "God is Unselfish Love", in *The Emptying God*, Ed. Cobb & Ives, pp. 119–120.

[63] Quoted in Moltmann, "God is Unselfish Love", p. 118.

[64] In this context, of course, "immanent" refers unfortunately to something completely different than the sense in which we have been using it thus far in discussing divine ontology.

[65] I am indebted to Ted Peters at this point for his conception of Barth opening the door to the economic Trinity becoming the immanent Trinity and subsequent theologians pressing the issue further. See Peters' *God as Trinity* (Louisville: Westminster/John Knox, 1993), p. 82.

[66] Eberhard Jüngel, *The Doctrine of the Trinity: God's Being is in Becoming*, p. 32.

[67] Barth, *CD*, II/1, p. 181. Quoted by Jüngel, p. 61.

[68] Though Jüngel does agree later with this identification in *God as the Mystery of the World* (Grand Rapids: Eerdmans, 1983), pp. 369–370.

[69] Karl Rahner, *The Trinity* (New York: Herder & Herder, 1970), pp. 21–22.

[70] Karl Rahner, *Theological Investigations, vol. IV: More Recent Writings*, trans. Kevin Smyth (London: Darton, Longman & Todd, 1966), p. 95.

[71] Rahner, *Theological Investigations, IV*, p. 95.

[72] Rahner, p. 96.

[73] Rahner writes, "When entering upon the doctrine of the Trinity, we need not hesitate to appeal to our own experience of Jesus and his Spirit in us as given in the history of salvation and faith .... [the Trinity] takes place in us" in *Theological Investigations, IV*, p. 98.

[74] Ibid., p. 97.

[75] Lewis Ford, "Process Trinitarianism" *Journal of the American Academy of Religion*, XLIII (1975), p. 207.

[76] Peters notes, "The essential problem of retrofitting process categories with biblical symbols is that the twoness of Whiteheadian metaphysics is difficult to reconcile with the threeness of trinitarian discourse. The God of Alfred North Whitehead's *Process and Reality* is binary or bipolar in that it has two natures: a primordial nature and a consequent nature. Similarly, Charles Harschorne was constantly occupied with reconciling the poles in the bipolar dialectic between God's absoluteness and relativity, between God's necessary existence and contingent actuality, between God's being both creator and recipient. Harshorne's insights into the double dimensions of the divine mark an achievement in philosophical theology. Yet he stops here, virtually ignoring the Trinitarian symbols of the absolute God of Israel who became relative in history. The Trinity is clearly not intrinsic to process metaphysics, so the value of the threefold revelation becomes doubtful", in Peters, p. 115.

[77] A. Guillaume, *Islam* (New York: Penguin, 1962), p. 131. This is not to say that Islam is completely without a developed doctrine of the free will of humanity – only that the tradition struggles to come up with a comprehensive vision of humanity's freedom in the context of overwhelming transcendence.

[78] Pannenberg notes that Augustine "had the courage not to resort to evasions at this point but even to emphasize the responsibility of the Creator for the development of things in his creation, though adding that God in his omniscience must have foreseen not only the fall of Adam but beyond it the further course of history until at last a descendant of Adam, with the help of God's grace, would overcome the devil." *Systematic Theology*, II, p. 264.

[79] See the extensive analysis in Barth, *CD*, II/1, pp. 569ff.

[80] Barth, using F. Diekamp, *Katholic Dogma*, vol. I, 1930, pp. 199–215.

[81] Thomas Aquinas, *Summa Theologica*, 83.1.3.

[82] Barth's quotation of J. Pohle's lively conclusion is worth noting: "Thomism is an imposing and strictly consistent view which gives impressive and lively expression to the omnipotence of God, His ability to initiate movement and cause all things, His supreme control and sovereignty. Yet the relentless out–working of its basic ideas in all spheres leads at specific points to a hardness and roughness which seriously disturb the harmonious integration of the architectonic plan ... Psychologically it produces gloom, includes to moral earnestness and gives rise to a conception of God that impels to fear. It is most congenial to characters of strong faith, while it can easily drive weak natures to despair. Therefore it is suitable only for lectures from the rostrum and cannot be utilized in rousing and inspiring preaching for ordinary Christian people. On the other hand, in Molinism we see softer and milder features, a high conception of God's loving providence, His merciful will to save, His grace that seeks men, His power to comply with feeble wills, coupled with His infinite lenience." J. Pohle, *Lehrbuch die Dogmatik*, vol. I, (1902), p. 203. Cf. K. Barth, *CD*, II/I, pp. 572ff.

[83] Cragg, *The Call of the Minaret*, pp. 306–307.

[84] Luther, *The Bondage of the Will*, in *Erasmus–Luther, Discourse on Free Will* (New York: Ungar, 1961), p. 112.

[85] Erasmus, *The Free Will*, in *Erasmus–Luther, Discourse on Free Will*, p. 32.

[86] Erasmus, p. 20. Weber notes that in the younger Melanchthon, Calvin, Peter Lombard, and in older Protestant Orthodoxy, the concept of "permission" was widely used (Weber, pp. 443–444).

[87] Erasmus, p. 59.

[88] Luther, *The Bondage of the Will*, p. 132.

[89] Weber, p. 444.

[90] See Abe, "Buddhism", *Our Religions*, p. 75.

[91] Phil Brown, "Buddhism & the Ecocrisis", (Buddha Darhma Education Association: Australia, 2003).

[92] Brown, "Buddhism & the Ecocrisis". Sandell quotes Lily de Silva: "The Buddhist admonition is to utilize nature in the same way as a bee collects pollen from a flower, neither polluting its beauty nor depleting its fragrance. Just as the bee manufactures honey out of pollen, so man should be able to find happiness without harming the natural world in which he lives." Klas Sandell, ed., *Buddhist Perspectives on the Ecocrisis* (Kandy, Sri Lanka: Buddhist Publication Society, 1987), p. 28.

[93] Klostermaier, p. 214.

[94] Ziauddin Sardar, *Towards an Islamic Theory of Environment* (Mansell, 1985), p. 225.

95 Nasr, p. 463. Cf. also David L. Johnston, *Toward Muslims and Christians as Joint Caretakers of Creation in a Postmodern World* (Fuller Theological Seminary Dissertation, 2001).

96 See Donald Worster, *The Wealth of Nature: Environmental History and the Ecological Imagination* (New York: Oxford, 1993), pp. 214–216.

97 Anderson, *On Being Human*, p. 63.

98 Grenz, *Theology for the Community of God*, p. 232.

99 Grenz, p. 232.

100 Anderson, p. 45.

101 Anderson, p. 218. Cf. also James Childs, *Christian Anthropology and Ethics* (Philadelphia: Fortress, 1978), p. 21.

102 Steven Bouma–Prediger, *For the Beauty of the Earth*, (Grand Rapids: Baker, 2001), pp. 96ff.

103 Bernhard Anderson, *From Creation to New Creation* (Minneapolis: Augsburg Fortress, 1994), p. 157.

104 Bouma–Prediger, *For the Beauty of the Earth*, pp. 73–74.

105 Wesley Granberg–Michaelson, *Ecology and Life* (Waco: Word, 1988), p. 35.

106 Bouma–Prediger, p. 104.

107 Ibid., p. 73.

108 Arnold Toynbee, "The Religious Background of the Present Environmental Crisis," in *Ecology and Religion in History*, ed. David & Eileen Spring (New York: Harper & Row, 1974), p. 147.

109 Toynbee, "The Religious Background of the Present Environmental Crisis," p. 148.

110 Roderick Nash, *The Rights of Nature: A History of Environmental Ethics* (Madison: University of Wisconsin, 1989), p. 90.

111 Lynn White, Jr., "The Historical Roots of our Ecologic Crisis," in *Science* 155 (March 10, 1967): 1203–1207.

112 For a response to White, see Wesley Granberg–Michaelson, *Ecology and Life*.

113 Bouma–Prediger, p. 74.

114 Ibid., p. 74.

115 Wendell Berry, *What Are People For?* (New York: North Point, 1990), p. 98.

116 Weber, p. 560.

117 Moltmann gives us a wonderful example of divine–centered human dominion: "As God's gifts, all his creatures are fundamentally eucharistic beings also; but the human being is able – and designated – to express the praise of all created things before God. In his own praise he acts as representative for the whole of creation. His thanksgiving, as it were, looses the dumb tongue of nature. It is here that the priestly dimension of his designation is to be found. So when in the 'creation' psalms thanks are offered *for* the sun and the light, *for* the heavens and the fertility of the earth, the human being is thanking God, not merely on his own behalf, but also in the name of heaven and earth and all created beings in them. Through human beings the sun and moon also glorify the Creator. Through human beings plants and animals adore the Creator too. That is why in the praise of creation the human being sings the cosmic liturgy, and through him the cosmos sings before its Creator the eternal song of creation. This is not meant anthropocentrically; for

in the community of creation 'everything that has breath' praises the Lord, and 'the heavens declare the glory of God' even without human beings – indeed as the representatives of human beings too, in their own way." (*God in Creation*, p. 71).

[118] Gerhard von Rad, *Genesis* (Philadelphia: Westminster, 1961), p. 58. Von Rad does not believe, however, that the mandate for dominion belongs to the description of humankind's image of God.

[119] Smart & Constantine, pp. 245–246.

Bibliography

Abe, Maseo. "Buddhism". *Our Religions.* Ed. Arvind Sharma. New York: Harper Collins, 1993.

_____. "Kenotic God and Dynamic Sunyata". *The Emptying God.* Ed. John B. Cobb & Christopher Ives. Maryknoll, NY: Orbis, 1990.

Adams, James Luther. "Introduction." in Ernst Troeltsch's *The Absoluteness of Christianity and the History of Religions.* Richmond: John Knox Press, 1971.

Al-Suhrawardy, Allama Sir Abdullah al-Ma'mun. *The Sayings of Muhammad.* New York: Carol Publishing Group, 1990.

Ali, Ahmed. *Al-Qur'ān.* Princeton: Princeton Unversity Press, 1988.

Anderson, Bernhard. *From Creation to New Creation.* Minneapolis: Augsburg Fortress, 1994.

Anderson, Ray. "Barth and a New Direction for Natural Theology". *Theology Beyond Christendom.* Ed. John Thompson. Allison Park, PA: Pickwick, 1986.

_____. *On Being Human.* Grand Rapids: Eerdmans, 1982.

Andrews, Allen A. *The Teachings Essential for Rebirth.* Tokyo: Sophia University, 1973.

Aquinas, Thomas. *De veritate* in *Truth; translated from the definitive Leonine text.* Chicago: H. Regnery Co., 1952-54.

_____. *Faith, Reason and Theology: questions I-IV of his Commentary on the De Trinitate of Boethius/ St. Thomas Aquinas.* Trans. A. Maurer. Toronto: Pontifical Institute of Medieval Studies, 1987.

_____. *Suma Contra Gentiles.* Notre Dame : University of Notre Dame Press, 1975.

_____. *Summa Theologiae. The Summa Theologica of St. Thomas Aquinas.* New York: Benziger Brothers, 1947.

Arnaldez, R. "Insān". *The Encyclopaedia of Islam, vol. IV.* Ed. E van Donzel, et.al. Leiden : E.J. Brill, 1978.

_____ . "Insān al-kāmil". *The Encyclopaedia of Islam, vol. IV.* Ed. E van Donzel, et.al. Leiden : E.J. Brill, 1978.

_____ . "Khalk". *The Encyclopaedia of Islam, vol. III.* Ed. B. Lewis, et.al. Leiden, E.J. Brill: 1971.

Augustine. *Against the Academics.* In *Ancient Christian Writers.* Ed. J. Quasten & J. Plumpe. Westminster, MD: The Newman Press, 1950.

_____. *On Christian Doctrine.* In *A Select Library of the Nicene and Post-Nicene Fathers of the Christian Church.* Ed. P. Schaff. Grand Rapids: Eerdmans, 1956.

_____. *The Correction of the Donatists.* In *A Select Library of the Nicene and Post-Nicene Fathers of the Christian Church.* Ed. P. Schaff. Grand Rapids: Eerdmans, 1956.

Barnes, Elizabeth. "Community, Narrative, and an Ecological Doctrine of Creation". *Theology without Foundations: Religious Practice and the Future of Theological Truth.* ed. Stanley Hauerwas, Nancey Murphy, Mark Nation. Nashville: Abingdon, 1994.

Barth, Karl. *Church Dogmatics.* Trans. Geoffrey Bromiley. Edinburgh: T. & T. Clark, 1975.

_____. *The Gottingen Dogmatics.* Ed. Hannelotte Reiffen. Trans. G. Bromiley. Grand Rapids: Eerdmans, 1991.

_____. *Learning Jesus Christ Through the Heidelberg Catechism.* Trans. Shirley C. Guthrie, Jr. Grand Rapids: Eerdmans, 1964.

Baum, Cunitz, & Reuss, Ed. *Corpus Reformatorum.* Brunswick, 1863-1900.

Bavinck, Herman. *Gereformeerde Dogmatiek.* Kampen: J.H. Bos, 1911.

Berkouwer, G.C. *The Providence of God.* Grand Rapids: Eerdmans, 1952.

Berry, Wendell. *What Are People For?* New York: North Point, 1990.

Bloesch, Donald G. *Holy Scripture: Revelation, Inspiration and Interpretation.* Downers Grove, IL: InterVarsity , 1994.

Bolle, Kees W. "Cosmology". *The Encyclopedia of Religion, vol. IV.* Ed. Mircea Eliade. New York: Macmillan, 1987.

Bonhoeffer, Dietrich. *Creation and Fall.* London: SCM Press, 1959.

_____. *Letters and Papers from Prison.* New York: Macmillan, 1953.

Boström, L. *The God of the Sages.* Stockholm, 1990.

Bouma-Prediger, Steven. *For the Beauty of the Earth.* Grand Rapids: Baker, 2001.

_____. *Toward a New Christian Theology of Nature.* Chicago: University of Chicago Dissertation, 1992.

Bouwsma, William J. *John Calvin: A Sixteenth Century Portrait.* New York: Oxford University Press, 1988.

Bray, Gerald. "Explaining Christianity to Pagans". *The Trinity in a Pluralistic Age.* Ed. Kevin J. Vanhoozer. Grand Rapids: Eerdmans, 1997.

Brockington, J.L. *Hinduism and Christianity.* New York: St. Martin's Press, 1992.

Brown, Colin. "Natural Theology". *New Dictionary of Theology.* ed. Sinclair B. Ferguson & David F. Wright. Downers Grove, Ill: InterVarsity Press, 1988.

Brown, Phil. "Buddhism and the Ecocrisis". Buddha Darhma Education Association: Australia, 2003.

Brunner, Emil. *Man in Revolt.* Philadelphia: Westminster, 1962.

_____. *Natural Theology.* London: Centenary Press. 1945.

_____. *Revelation and Reason.* Trans. Olive Wyon. Philadelphia: Westminster, 1946.

_____. *The Christian Doctrine of Creation and Redemption – Dogmatics, vol. II.* Trans. Olive Wyon. London: Lutterworth Press, 1952.

Bryce, Glendon E. *A Legacy of Wisdom.* New Jersey: Associated University Presses, 1979.

Buber, Martin. *I and Thou.* Trans. W. Kaufman. Edinburgh: T. & T. Clark, 1979.

Buccellati, G. "Wisdom and Not: The Case of Mesopotamia." *Journal of the American Oriental Society.* 101/1. 1981.

Calvin, John. *Institutes of the Christian Religion.* Tr. By Ford Lewis Battles. Ed. by John T. McNeill. *Library of Christian Classics.* Philadelphia: Westminster Press, 1960.

Camp, C.V. *Wisdom and the Feminine in the Book of Proverbs.* Sheffield, 1985.

Childs, James. *Christian Anthropology and Ethics.* Philadelphia: Fortress, 1978.

Chopp, Rebecca. *The Praxis of Suffering.* Maryknoll: Orbis, 1986.

Chung, Paul S. *Martin Luther and Buddhism.* Eugene, OR: Wipf & Stock, 2002.

Clifford, Richard J. & Collins, John J. ed. *Creation in the Biblical Traditions.* The Catholic Biblical Quarterly Monograph Series, 24, 1992.

Cobb, John B. *Beyond Dialogue: Towards a Mutual Transformation of Christianity and Buddhism.* Philadelphia: Fortress, 1982.

_____. "Beyond Pluralism". *Christian Uniqueness Reconsidered.* Ed. d'Costa. Maryknoll, NY: Orbis, 1990.

_____. & Ives, Christopher, ed. *The Emptying God: A Buddhist-Jewish-Christian Conversation.* Maryknoll, NY: Orbis, 1990.

Corduan, Winfried. *A Tapestry of Faiths.* InterVarsity Press, 2002.

Cragg, Kenneth. *The Call of the Minaret.* New York: Oxford University Press, 1956.

Crenshaw, J.L. "Prolegomenon". *Studies in Ancient Israelite Wisdom.* Ed. J.L. Crenshaw. New York: Ktav, 1976.

_____. *Old Testament Wisdom.* Louisville: Westminster John Knox, 1998.

D'Costa, Gavin. "Preface". *Christian Uniqueness Reconsidered.* Ed. Gavin d'Costa. Maryknoll, NY: Orbis, 1990.

_____. *Theology and Religious Pluralism.* Oxford: Basil Blackwell. 1986.

Dasgupta, Surendra Nath. *A History of Indian Philosophy*, vol. III. Cambridge: Cambridge University Press, 1952.

Davies, Brian. *The Thought of Thomas Aquinas.* Oxford: Clarendon, 1992.

Day, John. "Foreign Semitic influence on the wisdom of Israel and its appropriation in the book of Proverbs". *Wisdom in Ancient Israel.* Ed. John Day, Robert P. Gordon, & H.G.M. Williamson. Cambridge, 1995.

De Broglie, G. "La vraie notion thomiste des 'praeambula fidei,' " *Gregorianum* 34, 1953.

De La Vallēe Poussin, L. "Cosmology and Cosmogony (Buddhist)". *Encyclopedia of Religion and Ethics, vol. IV.* Ed. James Hastings. Edinburgh: T&T Clark, 1911.

Denzinger, H. *The Sources of Catholic Dogma.* Trans. R.J. Defarrari. 30th ed. St. Louis: Herder, 1957.

Descartes, Rene. *Discourse on Method.* 1637.

Diekamp, F. *Katholic Dogma, vol. I.* 1930.

Din, Khin Maung. "Some Problems and Possibilities for Burmese Christian Theology Today". *Christianity and the Religions of the East.* Ed. Richard W. Rousseau. Scranton, PA: Ridge Row Press, 1982.

DiNoia, Joseph. *The Diversity of Religions: A Christian Perspective.* Washington, D.C.: Catholic University Press, 1992.

Dupuis, Jacques. *Toward a Religious Theology of Religious Pluralism.* Maryknoll, NY: Orbis, 1997.

Eggeling, Julius, Trans. *Śatapatha Brāhmana*, according to the text of the *Madhyandina* School. Oxford: The Clarendon Press, 1889-1900.

Erasmus. *Erasmus-Luther, Discourse on Free Will.* New York: Ungar, 1961.

Feuerbach, Ludwig. *The Essence of Christianity.* Trans. George Eliot. New York: Harper and Brothers, 1957.

Ford, Lewis. "Process Trinitarianism". *Journal of the American Academy of Religion.* XLIII. 1975.

Gadamer, Hans Georg. *Truth and Method.* New York: Crossroad,1982.

Gardet, L. "Ibdā". *The Encyclopaedia of Islam, vol. III.* Ed. B. Lewis, et.al. Leiden, E.J. Brill: 1971.

Gese, H. *Lehre und Wirklichkeit in der alten Weishet.* Tübingen: Mohr/Siebeck, 1958.

Glasse, Cyril. *The Concise Encyclopedia of Islam.* San Francisco: Harper & Row, 1989.

Gogerly, Daniel John. *Ceylon Buddhism.* London: Kegan Paul, Trench, Trubner & Co., 1908.

Gransberg-Michaelson, Wesley. *Ecology and Life.* Waco: Word, 1988.

Grenz, Stanley. *Theology for the Community of God.* Nashville: Broadman & Holman, 1994.

Guillaume, A. *Islam.* New York: Penguin, 1962.

Guiness, Os. *Fit Bodies, Fat Minds.* Grand Rapids: Baker, 1994.

Gunton, Colin. *The One, the Three, and the Many: God, Creation and the Culture of Modernity.* Cambridge: Cambridge University Press, 1993.

_____. "The Spirit Moved Over the Face of the Waters: The Holy Spirit and the Created Order." *International Journal of Systematic Theology*, vol. 4, 2, July 2002.

_____. "The Trinity, Natural Theology, and a Theology of Nature". *The Trinity in a Pluralistic Age.* Ed. Kevin J. Vanhoozer. Grand Rapids: Eerdmans, 1997.

_____. *The Triune Creator.* Grand Rapids: Eerdmans, 1998.

Hadley, Judith M. *Evidence for a Hebrew Goddess: The Cult of Asherah in Ancient Israel and Judah.* Cambridge: Cambridge University Press, 2000.

_____. "Wisdom and the goddess." *Wisdom in Ancient Israel.* Ed. John Day, Robert P. Gordon, & H.G.M. Williamson. Cambridge, 1995.

Hart, Trevor. *Faith Thinking.* Downers Grove, Ill.: InterVarsity, 1995.

Heim, S. Mark. *The Depth of the Riches.* Grand Rapids: Eerdmans. 2001.

Heine, Heinrich. *The Romantic School.* New York: 1882.

Henry, Carl F.H. *God, Revelation and Authority.* Waco, Word, 1976-1983.

_____ "Postmodernism: The New Spectre" , in *The Challenge of Postmodernism: An Evangelical Engagement.* Wheaton: Victor Books, 1995.

Hick, John. *A Christian Theology of Religions*. Louisville: Westminster John Knox. 1995.

_____. *An Interpretation of Religion.* New York: Macmillan, 1989.

_____. *Death & Eternal Life*. New York: Harper & Row. 1976.

_____. *God Has Many Names*. Philadelphia: Westminster Press, 1982.

_____. "The Non-Absoluteness of Christianity". The Myth of Christian Uniqueness. Ed. John Hick & Paul F. Knitter. Maryknoll, NY: Orbis, 1987.

Holdrege, Barbara A. *Veda and Torah: Transcending the Textuality of Scripture*. New York: State University of New York Press, 1996.

Hooker, Roger. *Themes in Hinduism and Christianity.* Frankfurt: Verlag Peter Lang, 1989.

Jacobi, H. "Cosmogony & Cosmology (Indian)". *Encyclopedia of Religion and Ethics, vol. IV*. Ed. James Hastings. Edinburgh: T&T Clark, 1911.

Jenson, Robert. *Systematic Theology, vol. I.* New York: Oxford University Press, 1997.

Jewett, Paul. *God, Creation, and Revelation.* Grand Rapids: Eerdmans, 1991.

Johnston, David. *Toward Muslims and Christians as Joint Caretakers of Creation in a Postmodern World* (Fuller Theological Seminary Dissertation, 2001).

Johnston, Robert. "Rethinking Common Grace." *Grace Upon Grace*. Ed. Robert Johnston, L. Gregory Jones, Jonathon R. Wilson. Nashville: Abingdon, 1999.

Jones, Bruce, W. "From Gilgamesh to Qoheleth." *The Bible in the Light of Cuneiform Literature*. Ed. W.W. Hallo, B.W. Jones, & G.L. Mattingley. Lewiston: Edwin Mellon Press, 1990.

Jones, J.J., Trans. *The Mahavatsu*. London: Luzac, 1949.

Jüngel, Eberhard. *The Doctrine of the Trinity : God's Being is in Becoming*. Grand Rapids: Eerdmans, 1976.

_____. *God as the Mystery of the World.* Grand Rapids: Eerdmans, 1983.

Justin. *Apology*. In *The Ante-Nicene Fathers*, Vol. I. Ed. A. Roberts & J. Donaldson. Buffalo: The Christian Literature Pub. Co., 1885.

Kärkkäinen, Veli-Matti. *An Introduction to the Theology of Religions.* Downers Grove: InterVarsity Press, 2003.

_____. *Trinity and Religious Pluralism.* Ashgate, 2004.

Kaufman, Gordon D. "On the Meaning of 'God': Transcendence without Mythology." *Transcendence*. Ed. H.W. Richardson & D.R. Cutler. Boston: Beacon Press, 1969.

_____. "Religious Diversity and Religious Truth". *God, Truth and Reality.* Ed. Arvind Sharma. New York: St. Martin's Press, 1993.

Kayatz, C. *Studien zu Proverbian 1-9*. Neukirchen-Vluyn, 1966.

Khan, M. Muhsin, Trans. *Sahih al Bukhari: Arabic-English*. Beirut: Dar Al Arabic, 1985.

King, Winston L. *A Thousand Lives Away: Buddhism in Contemporary Burma*. Cambridge, MA: Harvard University Press, 1964.

Kloetzli, W. Randolph. *Buddhist Cosmology: Science and Theology in the Images of Motion and Light*. Delhi: Motalal Banarsidass, 1989.

_____. "Buddhist Cosmology". *The Encyclopedia of Religion, vol. IV*. Ed. Mircea Eliade. New York: Macmillan, 1987.

_____ "Hindu and Jain Cosmologies". *The Encyclopedia of Religion, vol. IV*. Ed. Mircea Eliade. New York: Macmillan, 1987.

Klostermaier, Klaus K. *A Survey of Hinduism*. Albany, NY: State University of New York Press, 1994.

Knight, Henry H. *A Future for Truth*. Nashville: Abingdon, 1997.

Knitter, Paul. "Horizons on Christianity's New dialogue With Buddhism". *Christianity and the Religions of the East*. Ed. Richard W. Rousseau. Scranton, PA: Ridge Row Press, 1982.

_____. *Introducing Theologies of Religions*. Maryknoll, NY: Orbis, 2003.

_____. "Introduction". *The Myth of Christian Uniqueness*. Ed. John Hick & Paul F. Knitter. Maryknoll, NY: Orbis, 1987.

Küng, Hans. "Christianity and World Religions: Dialogue with Islam". *Toward a Universal Theology of Religion*. Ed. Leonard Swidler. Maryknoll, NY: Orbis, 1987.

Lambert, W.G. *Babylonian Wisdom Literature*. Oxford: Clarendon, 1960.

_____. "Some new Babylonian wisdom literature." *Wisdom in Ancient Israel*. Ed. John Day, Robert P. Gordon, & H.G.M. Williamson. Cambridge, 1995.

Larson, Gerald J. *Classical Sāṃkyha: An Interpretation of Its History and Meaning*, 2nd ed. Santa Barbara, 1979.

Lasor, W.S., Hubbard, D., & Bush, F.W. *Old Testament Survey*. Grand Rapids: Eerdmans, 1982.

Latourette, Kenneth Scott. *A History of Christianity*, vol. 1. New York: Harper & Row, 1975.

Lefebure, Leo D. *The Buddha and the Christ*. Maryknoll, NY: Orbis, 1993.

Lewis, C.S. *The Weight of Glory*. New York: Macmillan Books, 1980.

Lindbeck, George. *The Nature of Doctrine*. Philadelphia: Westminster, 1984.

Long, Charles H. "Cosmogony". *The Encyclopedia of Religion, vol. IV*. Ed. Mircea Eliade. New York: Macmillan, 1987.

Luther, Martin. *Erasmus-Luther, Discourse on Free Will*. New York: Ungar, 1961.

Lyotard, Jean-François. *The Postmodern Explained*. trans. Don Barry. Minneapolis, 1993.

MacIntyre, Alasdair. *Three Rival Versions of Moral Enquiry*. Notre Dame: Notre Dame Press, 1989.

Martensen, H.L. *Christian Dogmatics*. Edinburgh, 1898.

Martinson, Paul. "Wisdom and Love as the Basis for Preaching in Buddhism and Chrstianity". *Christianity and the Religions of the East*. Ed. Richard W. Rousseau. Scranton, PA: Ridge Row Press, 1982.

McDermott, Gerald R. *Can Evangelicals Learn from World Religions?* Downers Grove, IL: Intervarsity. 2000.

McDermott, Timothy. *Thomas Aquinas: Selected Philosophical Writings*. New York: Oxford University Press, 1993.

McKim, Donald. *Westminster Dictionary of Theological Terms*. Louisville: Westminster, 1996.

Milbank, John. "The End of Dialogue". *Christian Uniqueness Reconsidered*. Ed. Gavin d'Costa. Maryknoll, NY: Orbis, 1990.

_____. *Theology & Social Theory*. Oxford: Oxford University Press, 1990.

Mitchell, E.K. "Cosmology and Cosmogony (Christian)". *Encyclopedia of Religion and Ethics, vol. IV*. Ed. James Hastings. Edinburgh: T&T Clark, 1911.

Moltmann, Jürgen. *The Church in the Power of the Spirit*. London: SCM Press, 1977.

_____. *The Crucified God*. New York: Harper & Row, 1974.

_____. *God in Creation*. Minneapolis: Fortress, 1993.

_____. "God is Unselfish Love". *The Emptying God*. Ed. John B. Cobb & Christopher Ives. Maryknoll, NY: Orbis, 1990.

_____. *The Trinity and the Kingdom of God*. London: SCM Press, 1981.

Murphy, Nancey. *Beyond Liberalism & Fundamentalism*. Valley Forge: Trinity Press, 1996.

Murphy, Roland. "The Personification of Wisdom." *Wisdom in Ancient Israel*. Ed. John Day, Robert P. Gordon, & H.G.M. Williamson. Cambridge, 1995.

_____. *The Tree of Life*. 2nd ed. Grand Rapids: Eerdmans, 1996.

Nāgārjuna, *Madhyamakakārikās*, in Frederick J. Streng, *Emptiness*. Nashville: Abingdon, 1967.

Nash, Roderick. *The Rights of Nature: A History of Environmental Ethics*. Madison: University of Wisconsin, 1989.

Nasr, Seyyed Hossein. "Islam". *Our Religions*. Ed. Arvind Sharma. New York: Harper Collins, 1993.

Netland, Harold. *Encountering Religious Pluralism*. Downers Grove, IL: Intervarsity. 2001.

Niebuhr, H. Richard. *Christ and Culture*. New York: Harper & Row, 1951.

Oberman, Heiko A. *Luther: Man between God and the Devil*. New Haven: Yale University Press, 1989.

Panikkar, Raimundo. *The Intra-Religious Dialogue*. Bangladore: Asian Trading Corporation, 1978.

_____. "The Jordan, the Tiber, and the Ganges". *The Myth of Christian Uniqueness*. Ed. John Hick & Paul F. Knitter. Maryknoll, NY: Orbis, 1987.

_____. *The Unknown Christ of Hinduism*. Rev. Ed. Maryknoll, NY: Orbis, 1979.

_____. *Trinity in World Religions*. CLS, Madras: 1970.

Pannenberg, Wolfhart. *Systematic Theology*, vol. I. Trans. Geoffrey Bromiley. Grand Rapids: Eerdmans, 1991.

_____. *Systematic Theology, vol. II.* Trans. Geoffrey Bromiley. Grand Rapids: Eerdmans, 1994.

Parker, T.H.L. *Calvin's Doctrine of the Knowledge of God.* Grand Rapids: Eerdmans, 1959.

Pegis, A.C. "Introduction." *Suma Contra Gentiles.* Notre Dame: University of Notre Dame Press, 1975.

Pelikan, J. Ed. *Luther's Works.* Saint Louis: Concordia, 1959.

Perdue, Leo G. *Wisdom & Creation.* Nashville: Abingdon, 1994.

Peters, Ted. *God as Trinity.* Louisville: Westminster/John Knox, 1993.

Pieris, Aloysius. "The Buddha and the Christ: Mediators of Liberation." *The Myth of Christian Uniqueness.* Ed. John Hick & Paul F. Knitter. Maryknoll, NY: Orbis, 1987.

Pinnock, Clark H. *Most Moved Mover.* Grand Rapids: Baker, 2001.

Pippen, Robert. *Modernism as Philosophical Problem.* Cambridge, Mass.: Basil Blackwell, 1991.

Pohle, J. *Lehrbuch die Dogmatik*, vol. I. 1902.

Portalie, Eugene. *A Guide to the Thought of Saint Augustine.* London: Burns & Oates. 1960.

Preuss, H.D. *Einfürung in die alttestamentlich Weisheitsliteratur.* Urban-Tashenbumlautcher 383. Stuttgart: Kohlhammer, 1987.

Pritchard, James B., Ed. *The Ancient Near East.* vol. 1. Princeton: Princeton University Press, 1958.

Quine, W.V.O. *From a Logical Point of View.* Cambridge: Harvard University Press, 1953.

Radhakrishnan, Sarvepalli & Moore, Charles. *A Sourcebook in Indian Philosophy.* Princeton, 1957.

Radhakrishnan, S. *The Principal Upaniṣads.* New York: Harper & Brothers, 1953.

Raghavachar, S. S., Ed. & Trans. *Vedārthasamgraha of Rāmānuja.* Mysore: Ramakrishna Ashrama, 1956.

Rahner, Karl. *Theological Investigations.* London, 1961-92.

_____. *The Trinity.* New York: Herder & Herder, 1970.

Ray, J.D. "Egyptian Wisdom Literature." *Wisdom in Ancient Israel.* Ed. John Day, Robert P. Gordon, & H.G.M. Williamson. Cambridge, 1995.

Ringgren, H. *Word and Wisdom.* Lund, 1947.

Robinson, John. *Truth is Two-Eyed.* Philadelphia: Fortress Press, 1979.

Samartha, Stanley J. "The Cross and the Rainbow: Christ in a Multireligious Culture." *The Myth of Christian Uniqueness.* Ed. John Hick & Paul F. Knitter. Maryknoll, NY: Orbis, 1987.

Sandell, Klas. *Buddhist Perspectives on the Ecocrisis.* Ed. Klas Sandell. Kandy, Sri Lanka: Buddhist Publication Society, 1987.

Sardar, Ziauddin. *Towards an Islamic Theory of Environment.* Mansell, 1985.

Schleiermacher, Friedrich. *On Religion: Speeches to Its Cultured Despisers.* Trans. J. Oman. New York : Harper, 1958.

_____. *The Christian Faith.* Edinburgh, 1928.

Schmid, H.H. "Creation, Righteousness and Salvation: 'Creation Theology' as the Broad Horizon of Biblical Theology". B.W. Anderson, ed. *Creation in the Old Testament.* London, 1984.

Schreiner, Susan. *The Theatre of His Glory.* Grand Rapids: Eerdmans, 1995.

Schwöbel, Christoph. "Particularity, Universality, and the Religions". *Christian Uniqueness Reconsidered.* Ed. Gavin d'Costa. Maryknoll, NY: Orbis, 1990.

Schuon, Frithjof. *Understanding Islam.* Trans. D.M. Matheson. New York: Roy, 1963. Seabury, 1978.

Sharma, Arvind. "Hinduism". *Our Religions.* Ed. Arvind Sharma. New York: Harper Collins, 1993.

_____. "The Puruṣāthas: An Axiological Exploration of Hinduism". *Journal of Religious Ethics*, 27:2.

Smart, Ninian and Constantine, Steven. *Christian Systematic Theology in a World Context.* Minneapolis: Fortress, 1991.

_____. *Buddhism and Christianity: Rivals and Allies.* Honolulu: University of Hawaii Press, 1993.

Smith, Gregory Bruce. *Neitzsche, Heidegger and the Transition to Postmodernity.* Chicago: Univeristy of Chicago Press, 1996.

Smith, M.S. *The Early History of God.* San Francisco, 1990.

Smith, W.C. "A Human View of Truth". *Truth and Dialogue.* Ed. J. Hick. London: Sheldon, 1974.

Sri Rahula, Walpola. *What the Buddha Taught.* New York: Grove Press, 1974.

Stout, Jeffrey. *The Flight from Authority.* Notre Dame: University of Notre Dame Press,1981.

Suchocki, Marjorie Hewett. "In Search of Justice". *The Myth of Christian Uniqueness.* Ed. J. Hick & Paul F. Knitter. Maryknoll, NY: Orbis, 1987.

Surin, Kenneth. "Towards a 'Materialist' Critique of 'Religious Pluralism'". *The Thomist*, 53:4, October 1989.

Suzuki, Daisetz Teitero. *Studies in the Lankāvatāra Sūtra.* London: Routledge & Kegan Paul Ltd., 1930.

Sweetman, J. Windrow. *Islam and Christian Theology, vol. II.* London: Lutterworth Press, 1947.

Tanner, Kathryn. "Respect for Other Religions: A Christian Antidote to Colonialist Discourse." *Modern Theology 9.* 1993.

Taylor, Mark C. *Erring.* Chicago: University of Chicago Press, 1984.

Tennet, T.C. *Christianity at the Religious Roundtable.* Baker, 2002.

Thiel, John E. *Nonfoundationalism.* Minneapolis: Fortress, 1994.

Thomas, Owen C., ed. *Attitudes Toward Other Religions.* New York: Harper & Row. 1969.

Tilley, Terrance. "In Favor of a 'Practical Theory of Religion'". *Theology without Foundations: Religious Practice and the Future of Theological Truth*, ed. Stanley Hauerwas, Nancey Murphy, Mark Nation. Nashville: Abingdon, 1994.

Tillich, Paul. *Biblical Religion and the Search for Ultimate Reality*. Chicago: University of Chicago Press, 1955.

_____. *Christianity and the Encounter of the World Religions*. New York: Columbia University Press, 1963.

_____. *Systematic Theology*. Chicago: University of Chicago Press. 1967.

_____. *The Future of Religions*. Ed. by Jerald C. Brauer. New York: Harper & Row, 1966.

Torrance, T.F. *Space, Time, and Incarnation*. New York, 1979.

Toulmin, Stephen. *Cosmopolis: The Hidden Agenda of Modernity*. New York: Macmillan, 1990.

Toynbee, Arnold. "The Religious Background of the Present Environmental Crisis." *Ecology and Religion in History*. Ed. David & Eileen Spring. New York: Harper & Row, 1974.

Tracy, David. *Blessed Rage for Order: The New Pluralism in Theology*. New York:

_____ *The Analogical Imagination*. New York: Crossroad, 1981.

_____ "Theological Method," *Christian Theology: An Introduction to Its Traditions and Tasks*. 2nd ed., ed. Peter C. Hodgson & Robert H. King. Philadelphia: Fortress, 1985.

Troeltsch, Ernst. *Christian Thought: Its History and Application*. London: University of London Press, 1923.

Van Huyssteen, J. Wentzel. *Essays in Postfoundationalist Theology*. Grand Rapids: Eerdmans, 1997.

Vanhoozer, Kevin J. "Does the Trinity Belong in a Theology of Religions?" *The Trinity in a Pluralistic Age*. Ed. Kevin J. Vanhoozer. Grand Rapids: Eerdmans 1997.

Vasabandu. *Abihiddarmakosbhasyam*. Rendered by L. de La Valleé Poussin. Eng. Trans. Leo M. Pruden. Berkeley, CA: Asian Humanities Press, 1988.

Volf, Miroslav. *Exclusion and Embrace*. Nashville: Abingdon, 1996.

Von Rad, Gerhard. *Genesis*. Philadelphia: Westminster, 1961.

_____. *Old Testament Theology, vol. 1*. New York: Harper & Row, 1962.

_____. *Wisdom in Israel*. Nashville: Abingdon, 1972.

Vos, Arvin. *Aquinas, Calvin & Contemporary Protestant Thought*. Washington, D.C.: Christian University Press, 1985.

Wallace, Ronald. *Calvin's Doctrine of the Word and Sacrament*. Grand Rapids: Eerdmans, 1957.

Ward, Graham. "Postmodern Theology" in *The Modern Theologians*, 2nd ed. ed. David F. Ford. Cambridge, Mass.: Blackwell, 1997.

Watt, W. Montgomery. *Islam and Christianity Today*. London: Routledge & Kegan Paul, 1983.

Weber, Otto. *Foundations of Dogmatics, vol. 1.* Trans. D. Guder. Grand Rapids: Eerdmans, 1981.

Westermann, Claus. *Genesis.* Minneapolis: Augsburg Fortress, 1992.

Whaling, Frank. "The Trinity and the Structure of Religion Life". *Christianity and the Religions of the East.* Ed. Richard W. Rousseau. Scranton, PA: Ridge Row Press, 1982.

White Jr., Lynn. "The Historical Roots of our Ecologic Crisis". *Science.* 155 (March 10, 1967).

Whybray, R.N. *Ecclesiastes.* Grand Rapids & London, 1989.

Wilson, H.H. Trans.and Ed. *The Vishnu Purāṇa* (1840). 2nd ed. Calcutta, 1961.

Wolleb, J. *Compendium theologiae Christianae.* In *Reformed Dogmatics.* Ed. J.W. Beardslee III. New York Oxford University Press, 1965.

Wolterstorff, Nicholas. *Reason Within the Bounds of Religion.* Grand Rapids: Eerdmans, 1976.

Woodbury, J. Dudley. "Different Diagnoses of the Human Condition". *Muslims and Christians on the Emmaus Road.* Ed. J. Dudley Woodbury. Pasadena: Zwemer Institute of Christian Studies, 1989.

Worster, Donald. *The Wealth of Nature: Environmental History and the Ecological Imagination.* New York: Oxford, 1993.

Würthwein, Ernest. "Egyptian Wisdom and the Old Testament." *Studies in Ancient Israelite Wisdom.* Ed. J.L. Crenshaw. New York: Ktav, 1976.

Yagi, Seiichi. "'I' in the Words of Jesus." *The Myth of Christian Uniqueness.* Ed. John Hick & Paul F. Knitter. Maryknoll, NY: Orbis, 1987.

Yong, Amos. *Beyond the Impasse.* Grand Rapids: Baker, 2003.

Zaehner, R.C. *Hindu Scriptures.* London: Dent, 1966.

Zimmerli, W. "Concerning the Structure of Old Testament Wisdom." *Studies in Ancient Israelite Wisdom.* Trans. Brian W. Kovacs. Ed. J.L. Crenshaw. New York: Ktav, 1976.

_____. "The Place and Limit of the Wisdom in the Framework of the Old Testament Theology." *Scottish Journal of Theology.* 17. 1964.

INDEX

A

B

C

D

E

F

G

H